SpringerBriefs in Mathematical Physics

Volume 35

SpringerBriefs are characterized in general by their size (50–125 pages) and fast production time (2–3 months compared to 6 months for a monograph).
Briefs are available in print but are intended as a primarily electronic publication to be included in Springer's e-book package.

Typical works might include:

- An extended survey of a field
- A link between new research papers published in journal articles
- A presentation of core concepts that doctoral students must understand in order to make independent contributions
- Lecture notes making a specialist topic accessible for non-specialist readers.

SpringerBriefs in Mathematical Physics showcase, in a compact format, topics of current relevance in the field of mathematical physics. Published titles will encompass all areas of theoretical and mathematical physics. This series is intended for mathematicians, physicists, and other scientists, as well as doctoral students in related areas.

More information about this series at http://www.springer.com/series/11953

Fumio Hiroshima

Ground States of Quantum Field Models

Perturbation of Embedded Eigenvalues

Springer

Fumio Hiroshima
Faculty of Mathematics
Kyushu University
Fukuoka, Japan

ISSN 2197-1757 ISSN 2197-1765 (electronic)
SpringerBriefs in Mathematical Physics
ISBN 978-981-32-9304-5 ISBN 978-981-32-9305-2 (eBook)
https://doi.org/10.1007/978-981-32-9305-2

This Springer imprint is published by the registered company Springer Nature Singapore Pte Ltd.
The registered company address is: 152 Beach Road, #21-01/04 Gateway East, Singapore 189721, Singapore

To my family and friends

Preface

The book aims to give a survey of the existence of the ground state in quantum field theory from an operator theoretical point of view. In particular, I focus on investigating quantum interaction systems between non-relativistic quantum mechanical matters and quantum fields. Mathematically the energy Hamiltonian of a quantum interaction system can be realized as a selfadjoint operator H acting in a Hilbert space $\mathscr{H}$. The selfadjoint operators associated with quantum interaction systems we consider are Schrödinger-type operators coupled to quantum fields. The problem reviewed in this book can be reduced to studying perturbations of eigenvalues embedded in the continuous spectrum of the selfadjoint operators. In particular, stability and instability of the lowest eigenvalue of the selfadjoint operators are the main subjects.

If the bottom of the spectrum of a selfadjoint operator is an eigenvalue, we say that there exists a ground state of the selfadjoint operator or the selfadjoint operator has a ground state.

In quantum field theory, there appears *infrared problem* owing to the vanishing mass of bosons. On account of infrared problem, it is not trivial to show the existence of the ground state of H. Nevertheless, several versions of proofs of the existence of the ground state of H have been developed since the end of the twentieth century, and huge amount of related papers have been published to date. It is, however, impossible to make a complete list of them in this book. The reader may find descriptions and references in, e.g., Arai [4], Gustafson and Sigal [22], and Spohn [55].

A new point of view on the existence of the ground state emerged around 1995 with works by Bach, Fröhlich, and Sigal [6–9] in which the existence of the ground state of the so-called Pauli-Fierz model in non-relativistic quantum electrodynamics was shown, and renormalization group method based on Feshbach maps was introduced to study resonances. Simultaneously and independently, Arai and Hirokawa [5] also proved the existence of the ground state of generalized spin-boson models. In 1999, Spohn [54] succeeded in proving the existence of the ground state of the so-called Nelson model by a functional integral method and simultaneously Gérard [17] by compact operator arguments. In particular, [17] considers an abstract model including the Nelson model. In [17, 54], the existence of the ground state is proven, no matter how large a coupling constant is.

In this period, I was spending brilliant days as a postdoctoral student (JSPS fellow) in Universität Bonn (February 1998–January 1999) and Technische Universität München (February 1999–February 2000). In the postdoctoral period in Germany, I was inspired by [5, 9, 17, 54], and started studying spectral analysis of Hamiltonians related to quantum field theory and collaborated with Herbert Spohn in München. I attended "Les Houches Summer School of Physics 1998" in Chamonix where my research was encouraged by an encounter with a lot of distinguished researchers coming from all over the world and tried to climb up the Mont Blanc in holidays during the summer school. Several participants of the summer school asked me how the top of the Mont Blanc was, but nobody unfortunately asked me my own scientific research. In the autumn of 1998, I gave a talk in a mathematical physics seminar in Technische Universität Berlin, where after my talk Volker Bach taught me to be able to avoid infrared divergence in non-relativistic quantum electrodynamics. This surprised me very much. I was also invited to an international conference held in Lille in France in 1999, where Christian Gérard gave a talk on the existence of the ground state just before my talk. It was too impressive to talk for me. In 2001, Griesemer, Lieb and Loss [21] and Lieb and Loss [43] proved the existence of the ground state of the Pauli-Fierz model under the so-called binding condition, in which a compact embedding of a Sobolev space was applied.

As was stated above, spectral analysis of the Pauli-Fierz mode, the Nelson model, and related models was progressed very quickly at the end of the twentieth century. Without a break in the twenty-first century, a lot of important papers concerned with non-relativistic quantum electrodynamics, etc. were also produced at an increasing tempo. I may say that results accumulate and eventually reach a critical state nowadays.

In July 2015, I attended ICMP2015 held in Santiago in Chile and just after my short talk concerned with the semi-relativistic Pauli-Fierz Hamiltonian, a staff of Springer invited me to publish a book about ground state "not about functional integration" which is my scientific hometown. I first intended to make my own results as a book, but I thought my own results were insignificant extensions of results established in [5, 9, 17, 21, 54] mentioned above. What then one really needs is a single source that combines fundamental results on the existence of ground states obtained in the end of the twentieth century with a guide to how to use them to other models reader may consider, because I could not find satisfactory discussions in literature. Then I decided to write up my own account of the existence of ground states.

However, any book is a compromise between deadlines and my striving for perfections. Without deadlines no book would be published, this means a final version which is perfect exists only in my mind. Due to time and space constraints, I did not include fundamental mathematical tools but the book is written as possible as self-consistently. The present book comprises material which can partly be found in other books and partly in research articles, and which for the first time is exposed from a unified point of view.

The book consists of six chapters. Chapter 1 is overview of models studied in this book without rigor. Here, we introduce several strategies on how to show the existence of ground states. Chapter 2 is devoted to a minimal introduction of tools I used in this book. Compact operators, compact sets in L^p space, compact embeddings of Sobolev spaces, and boson Fock space are reviewed. Chapter 3 is devoted to studying the Pauli-Fierz model and Chap. 4 the Nelson model with/without cutoffs. In both chapters, we prove the existence of ground states by applications of compactness. In Chap. 5, I study spin-boson model by path measures. Finally in Chap. 6, I overview enhanced bindings for the Pauli-Fierz model and the Nelson model.

Fukuoka, Japan Fumio Hiroshima

Acknowledgements I would like to very heartily thank all those who helped me for the product of book. I am grateful for helpful comments and discussions from Thomas Norman Dam, Takeru Hidaka, Masao Hirokawa, József Lőrinczi, Oliver Matte, Jacob Schach Møller, Itaru Sasaki, Akito Suzuki, and Herbert Spohn. I am financially supported by JSPS KAKENHI 16H03942, CREST JPMJCR14D6, and JSPS open partnership joint research with Denmark 1007867.

Contents

Notations

a	Annihilation operator
$a^\dagger$	Creation operator
$(a\Phi)^{(n)}(k)$	$\sqrt{n}\Phi^{(n)}(k,\cdot)$
A_μ	Quantized radiation field
α_c	Critical coupling constant
α	Coupling constant of $H_{\mathrm{PF}}^{\mathrm{dip}}$
$[A,B]$	Commutator: $AB-BA$
$B_c(X)$	Compact operators on X
$B(X,Y)$	Bounded operators from X to Y
$B(X)$	$B(X,X)$
$B_1(X)$	Trace classes on X
$B_2(X)$	Hilbert-Schmidt operators on X
$(B_t)_{t\geq 0}$	Brownian motion
$\beta(k)$	$(\omega(k)+\|k\|^2/2m)^{-1}$
$C^k(U)$	Continuously k times differentiable functions on U
$C_0^\infty(\mathbb{R}^d)$	Infinite times differentiable functions on $\mathbb{R}^d$ with compact support
$C_u^{\gamma,k}(U)$	Hölder space
$\mathrm{d}\Gamma(h)$	Differential second quantization
$D(T)$	The domain of T
Df	$Df=(\partial_{x_1}f,\ldots,\partial_{x_d}f)$
D^α	$\partial_{x_1}^{\alpha_1}\ldots\partial_{x_d}^{\alpha_d}$
$e(k,j)$	Polarization vector
$E(H)$	Bottom of the spectrum of H
E_R	$\inf_{F\in Q(\mathfrak{q}),\tilde{\chi}_R F\neq 0}\mathfrak{q}(\tilde{\chi}_R F,\tilde{\chi}_R F)/(\tilde{\chi}_R F,\tilde{\chi}_R F)$
$E_{0,R}$	$\inf_{F\in Q(\mathfrak{q}_0),\tilde{\chi}_R F\neq 0}\mathfrak{q}_0(\tilde{\chi}_R F,\tilde{\chi}_R F)/(\tilde{\chi}_R F,\tilde{\chi}_R F)$
$E_{\kappa\lambda,R}$	$\inf_{F\in H_0^{1/2},\tilde{\chi}_R F\neq 0}B_{\kappa\lambda}(\tilde{\chi}_R F,\tilde{\chi}_R F)/(\tilde{\chi}_R F,\tilde{\chi}_R F)$
E_σ	$\inf\mathrm{sp}(H_\sigma)$
$E_{\kappa\lambda}$	$\inf\mathrm{sp}(H_{\kappa\lambda})$

$E_{\kappa\lambda,0}$	$\inf\mathrm{sp}(H_{\kappa\lambda,V=0})$
$\mathbb{E}[\cdots]$	Expectation of $[\cdots]$
E_O	Spectral projection
E_{el}	$\inf\mathrm{sp}(H_{\mathrm{el}}) = -\|g\|^4/(32\pi^2)$
$\hat{f}$	Fourier transform of f
$\check{f}$	Inverse Fourier transform of f
$\mathscr{F}$	Boson Fock space
g, g_j	Coupling constants of H_{N}
$\Gamma(T)$	Second quantization
H_{p}	Schrödinger operator: $-\Delta/2+V$
h_{eff}	Schrödinger operator with effective mass: $-\Delta/2m_{\mathrm{eff}}+V$
$\mathscr{H}$	Total Hilbert space: Pauli-Fierz model $L^2(\mathbb{R}^3)\otimes\mathscr{F}(\mathscr{H}_{\mathrm{PF}})$; Nelson model $L^2(\mathbb{R}^d)\otimes\mathscr{F}(L^2(\mathbb{R}^d))$, Spin-Boson model $\mathbb{C}^2\otimes\mathscr{F}(L^2(\mathbb{R}^d))$.
$\mathscr{H}_{\mathrm{PF}}$	$L^2(\mathbb{R}^3\times\{1,2\})$
H_{rad}	Free field Hamiltonian of H_{PF}:$\mathrm{d}\Gamma(\omega)$
H_{PF}	Pauli-Fierz Hamiltonian
$H_{\mathrm{PF}}^{\mathrm{dip}}$	Pauli-Fierz Hamiltonian with dipole approximation
$H_{\mathrm{PF},0}$	H_{PF} with zero coupling: $H_{\mathrm{p}}\otimes\mathbb{1}+\mathbb{1}\otimes H_{\mathrm{rad}}$
H_{f}	Free field Hamiltonian of H_{N} and H_{SB}:$\mathrm{d}\Gamma(\omega)$
H_{N}	Nelson Hamiltonian
$H_{\mathrm{N},0}$	H_{N} with zero coupling: $H_{\mathrm{p}}\otimes\mathbb{1}+\mathbb{1}\otimes H_{\mathrm{f}}$
H_σ	Infrared cutoff Nelson Hamiltonian: $H_0+\phi_\sigma$
$\tilde{H}_\sigma$	$H_{\mathrm{p}}+\mathrm{d}\Gamma(\tilde{\omega}_\sigma)+\phi_\sigma$
$\tilde{H}_{0,\sigma}$	$H_{\mathrm{p}}+\mathrm{d}\Gamma(\tilde{\omega}_\sigma)$
$\tilde{H}_\sigma^{\mathrm{ex}}$	$\tilde{H}_\sigma\otimes\mathbb{1}+\mathbb{1}\otimes\mathrm{d}\Gamma(\tilde{\omega}_\sigma)$
$\tilde{H}_{0,\sigma}^{\mathrm{ex}}$	$\tilde{H}_{0,\sigma}\otimes\mathbb{1}+\mathbb{1}\otimes\mathrm{d}\Gamma(\tilde{\omega}_\sigma)$
$H_{\kappa\lambda}$	Gross transformed Nelson Hamiltonian with cutoff $0\le\kappa<\lambda\le\infty$
H_0	$H_{\kappa\lambda}$ with zero coupling:$(H_{\mathrm{el}}-E_{\mathrm{el}})\otimes\mathbb{1}+\mathbb{1}\otimes H_{\mathrm{f}}$
H_{SB}	Spin-Boson Hamiltonian
$\tilde{H}_{\mathrm{SB}}$	$UH_{\mathrm{SB}}U^*$
$H_{\mathrm{SB},0}$	H_{SB} with zero coupling: $\varepsilon\sigma_z\otimes\mathbb{1}+\mathbb{1}\otimes H_{\mathrm{f}}$
I	$\int_{\mathbb{R}^d}\|\hat{\varphi}(k)\|^2/\omega(k)^3dk$
K_E	$R_E^*R_E$
$\kappa_j(k)$	$(H_{\mathrm{PF}}-E+\omega(k))^{-1}D_j\Psi_\nu$
$l_{\mathrm{loc}}(\mathbb{R}^d)$	Pair of smooth partition functions $\{j_1,j_2\}$
m_c	Critical mass given by $\\|K_0\\|^{-1}$
m_{eff}	Effective mass
N	Number operator
$(N_t)_{t\ge0}$	Poisson process
ν	Artificial mass for H_{PF}

$o(R^k)$	Function such that $\lim_{R\to\infty} o(R^k)/R^k = 0$
$o(R^0)$	Function such that $\lim_{R\to\infty} o(R) = 0$
ω	Dispersion relation: $\sqrt{\lvert k\rvert^2 + \nu^2}$ or $\lvert k\rvert$
$\tilde{\omega}_\sigma$	Smooth infrared cutoff dispersion relation: $\tilde{\omega}_\sigma(k)\begin{cases} \geq \sigma/2, & \lvert k\rvert < \sigma, \\ = \omega(k), & \lvert k\rvert \geq \sigma. \end{cases}$
Ω	Fock vacuum
Ω_{SB}	$\begin{pmatrix} 1 \\ 1 \end{pmatrix} \otimes \Omega$
p^*	Conjugate of p: $dp/(d-p)$
P_Ω	Projection to $\{\alpha\Omega \mid \alpha \in \mathbb{C}\}$
$\wp$	Cutoff function of A_μ
$\hat{\varphi}$	Cutoff function of H_{PF}, H_{N} and H_{SB}
ϕ_σ	$\phi(\rho_\sigma)$
Ψ_{g}	Ground state of H_{PF}, H_{N} and H_{SB}
$\Psi_{\kappa\lambda}$	Ground state of $H_{\kappa\lambda}$
Ψ_σ	Ground state of H_σ with infrared cutoff σ
Ψ_ν	Ground state of H_{PF} with positive mass ν
q	Quadratic form associated with H_{PF}
q_0	q with $V = 0$
q_A	Quadratic form associated with selfadjoint operator A
$Q(\mathrm{q})$	Form domain of q
$B_{\kappa\lambda}$	Quadratic form associated with $H_{\kappa\lambda}$
R_E	Birman-Schwinger operator: $(-\Delta/2 - E)^{-1/2}\lvert V\rvert^{1/2}$
$\rho(x)$	$\hat{\varphi}(k)e^{-ikx}/\sqrt{\omega(k)}$
ρ_σ	$\rho \mathbb{1}_{\omega \geq \sigma}$
σ	Infrared cutoff for H_{N}
$\sigma_{\varepsilon t}$	$\sigma(-1)^{N_{\varepsilon t}}$
S_n	Symmetrizer on $\otimes^n \mathscr{H}$
$\mathrm{sp}(h)$	Spectrum of h
$\mathrm{sp}_{\mathrm{disc}}(h)$	Discrete spectrum of h
Σ	Ionization energy of H_{PF}
$\Sigma_{\kappa\lambda}$	Ionization energy of $H_{\kappa\lambda}$
Σ_V	The lowest two cluster threshold
$\mathscr{S}(\mathbb{R}^d)$	Schwartz space on $\mathbb{R}^d$
$\mathscr{S}'(\mathbb{R}^d)$	Schwartz distributions on $\mathbb{R}^d$
$s-\lim$	Strong limit
$T_{g,j}$	$T_{g,j}f = -\int_{\mathbb{R}^3} f(k)\kappa_j(k)dk$
$\mathscr{U}_{\mathrm{c}}$	Canonical unitary mapping: $\mathscr{F} \to \mathscr{F}_1 \otimes \mathscr{F}_2$
V_{Kato}	Class of external potentials
$V_{\mathrm{Kato}}^{\mathrm{q}}$	Class of external potentials for form
V_{eff}	Effective potential

$W^{k,p}(U)$	Sobolev space
$w-\lim$	Weak limit
$\langle x \rangle$	Japanese bracket: $\sqrt{1+\|x\|^2}$
$\mathbb{Z}_2$	$\{-1,1\}$

Chapter 1
Introduction

1.1 Quantum Field Theory and Ground States

Quantum field theory was originally thought to be the quantum theory of classical fields. Nowadays it turns out to be a theoretical framework that combines classical field theory, special relativity, and quantum mechanics. It is used to construct physical models of subatomic particles in particle physics and quasiparticles in condensed matter physics, etc.

When quantum mechanics was discovered in the middle of 1920s, physicists already knew about various classical fields, notably the electromagnetic field. In 1926 by Born, Heisenberg and Jordan [10] the formalism of matrix mechanics developed by Heisenberg [25] was extended to systems having arbitrarily many degrees of freedom [10, p. 338]. They also discussed canonical transformations, perturbation theory, angular momentum, eigenvalues and eigenvectors, where they gave a formula for the electromagnetic field [10, p. 375] as a Fourier transform and used canonical commutation relations to identify coefficients in this Fourier transform as operators that create and annihilate photons. In addition to the formalism, [10] also contains the earliest discussion of a quantum field theory. It was soon after [10] that one could use quantum field theory for everything, not just for the electromagnetic field. The idea of quantum field theory is that quantum fields are the basic ingredients of the universe, and particles are just bundles of energy and momentum of the fields.

In this book we discuss a system of non-relativistic quantum mechanical matters coupled to quantum fields from purely mathematical point of view. Non-relativistic quantum mechanical matters treated in this book are electrons, nucleons and two-level atoms, etc. It is assumed that these particles are non-relativistic and spinless particles. We assume that these quantum mechanical matters are governed by Schrödinger operators. We study three models.

(1) the Pauli–Fierz model with/without dipole approximation,
(2) the Nelson model with/without cutoffs,
(3) spin-boson model.

F. Hiroshima, *Ground States of Quantum Field Models*,
SpringerBriefs in Mathematical Physics,
https://doi.org/10.1007/978-981-32-9305-2_1

The Pauli–Fierz model was introduced by Pauli and Fierz [50] in 1938, which describes a system of non-relativistic and spinless electrons minimally coupled to a quantized radiation field. The Nelson model was introduced by Nelson [49] in 1964, which describes non-relativistic and spinless nucleons linearly coupled to a scalar meson field. Finally spin-boson model describes a two-level atom coupled to a scalar bose field which was investigated in e.g. [41]. Hamiltonians of these models can be realized as selfadjoint operators acting in some infinite dimensional Hilbert spaces which are the tensor product $\mathscr{H}_1 \otimes \mathscr{H}_2$ of two Hilbert spaces $\mathscr{H}_1$ and $\mathscr{H}_2$. Here $\mathscr{H}_1$ describes the state space of non-relativistic quantum mechanical matters and $\mathscr{H}_2$ a boson Fock space. The ultimate goal is to specify the spectra of these selfadjoint operators.

Definition 1.1 (*Ground state and ground state energy*) Let K be a selfadjoint operator. The bottom of the spectrum $E = \inf \mathrm{sp}(K)$ is called ground state energy and an eigenvector ψ (if it exists) associated with E is called ground state:

$$K\psi = E\psi.$$

We are concerned with the bottom of the spectra of the Pauli–Fierz Hamiltonian, the Nelson Hamiltonian and spin-boson Hamiltonian. Stability and instability of ground states are main subjects of concern in this book.

Before making statements mathematically sharp, in Sects. 1.2–1.7 below we shall overview models and problems addressed in this book without rigor for readers convenience. If one is familiar with the Pauli–Fierz model etc, one can skip Sects. 1.2–1.7. The rigorous discussions start from Chap. 2.

1.2 The Pauli–Fierz Model

The Pauli–Fierz Hamiltonian describes the minimal interaction between electrons and a quantized radiation field, where electrons are treated as non-relativistic and spinless quantum mechanical matters and the elementary particle counterpart of the quantized radiation field is light quanta "photons". Hence the Pauli–Fierz Hamiltonian is defined by a Schrödinger operator coupled to a quantized radiation field A. Thus the Pauli–Fierz model is usually said to be a model in non-relativistic quantum electrodynamics. In Chap. 3 we discuss the existence of the ground state of the Pauli–Fierz Hamiltonian.

Consider the Schrödinger operator of the hydrogen like atom,

$$H_{\mathrm{hyd}} = -\frac{1}{2}\Delta - \frac{a}{|x|}, \quad a > 0,$$

acting in $L^2(\mathbb{R}^3)$. The Hamiltonian of the state space of photons is denoted by H_{rad}. $H_{\mathrm{PF},0}$ is defined by

Fig. 1.1 Spectrum of $H_{\mathrm{PF},0}$

$$H_{\mathrm{PF},0} = H_{\mathrm{hyd}} \otimes \mathbb{1} + \mathbb{1} \otimes H_{\mathrm{rad}}.$$

The spectrum of the hydrogen like atom is disjoint union of negative eigenvalues and essential spectrum:

$$\mathrm{sp}(H_{\mathrm{hyd}}) = \{-\frac{a^2}{2n^2}\}_{n=1}^{\infty} \cup [0, \infty) \tag{1.1}$$

and that of H_{rad} is

$$\mathrm{sp}(H_{\mathrm{rad}}) = [0, \infty) \tag{1.2}$$

with the unique point spectrum $\{0\}$, i.e., the bottom of the spectrum of H_{rad} is an eigenvalue lying in the essential spectrum. The spectrum of $H_{\mathrm{PF},0}$ is[1]

$$\mathrm{sp}(H_{\mathrm{PF},0}) = \{a + b | a \in \mathrm{sp}(H_{\mathrm{hyd}}), b \in \mathrm{sp}(H_{\mathrm{rad}})\} = [-\frac{a^2}{2}, \infty)$$

and all eigenvalues $\{-a^2/2n^2\}_{n=1}^{\infty}$ are *embedded* in the continuous spectrum. See Fig. 1.1.

The full Pauli–Fierz Hamiltonian is defined by imposing the minimal coupling of a quantized radiation field on $H_{\mathrm{PF},0}$:

$$-\Delta_x \otimes \mathbb{1} = (-i\nabla_x)^2 \otimes \mathbb{1} \to (-i\nabla_x \otimes \mathbb{1} - \alpha A(x))^2,$$

where $A(x) = A_{\hat{\varphi}}(x) = (A_1(x), A_2(x), A_3(x))$ is a quantized radiation field with cutoff function $\hat{\varphi}$ and $\alpha \in \mathbb{R}$ a coupling constant. The Pauli–Fierz Hamiltonian is defined by

$$H_{\mathrm{PF}} = \frac{1}{2}(-i\nabla \otimes \mathbb{1} - \alpha A(x))^2 - \frac{a}{|x|} \otimes \mathbb{1} + \mathbb{1} \otimes H_{\mathrm{rad}}.$$

In Chap. 3 we give a rigorous definition of H_{PF}, where $-a/|x|$ is replaced by a general external potential V. We are interested in specifying the spectrum of H_{PF} for $\alpha \neq 0$. In particular the behavior of the lowest eigenvalue is the main subject in this book.

The difficulty in establishing the behavior comes from the fact that the bottom of the spectrum lies in the essential spectrum, not below it, as is the case

[1]In general $\mathrm{sp}(X \otimes \mathbb{1} + \mathbb{1} \otimes Y) = \{x + y | x \in \mathrm{sp}(X), y \in \mathrm{sp}(Y)\}$.

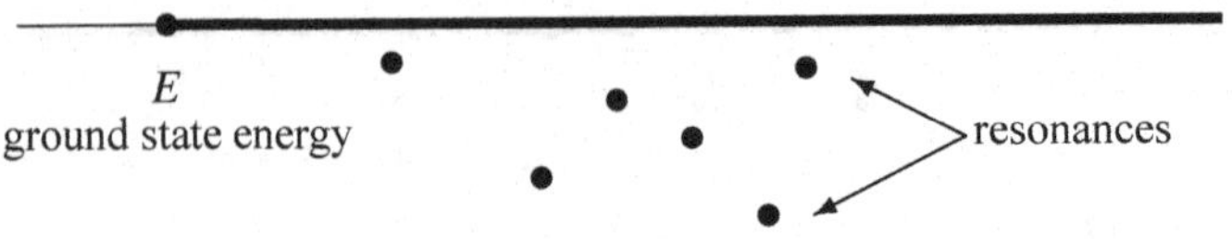

Fig. 1.2 Spectrum of H_{PF}

for usual Schrödinger operators. Let us consider Schrödinger operator $-\Delta/2 + V$ with external potential V which satisfies $V \in L^\infty(\mathbb{R}^3)$ and $|V(x)| \to 0$ as $|x| \to \infty$. This assumption yields that V is relatively compact with respect to $-\Delta/2$; $V(-\Delta/2 + \lambda)^{-1}$ is compact for any $\lambda > 0$ and the essential spectrum of $-\Delta/2 + V$ is $[0, \infty)$. Let e be the bottom of the spectrum of $-\Delta/2 + V$ and e_0 that of $-\Delta/2$. e_0 is equal to the bottom of the essential spectrum of $-\Delta/2 + V$. Then $e_0 = 0$. If

$$e < e_0,$$

then e is discrete and we can conclude that $-\Delta/2 + V$ has a ground state.

Consider H_{PF}. Similar to $-\Delta/2 + V$, we denote the bottom of the spectrum of H_{PF} and H_{PF} with $a = 0$ by E and E_0, respectively. In the case of the Pauli–Fierz Hamiltonian, despite inequality

$$E < E_0,$$

E lies in the bottom of the essential spectrum, and it is unclear that H_{PF} admits a ground state.

Since photon is massless,[2] the energy of a single photon with momentum $k \in \mathbb{R}^3$ is given by $|k|$. Hence a physical reason why E lies in the bottom of the essential spectrum is that we can always add arbitrarily many and arbitrarily little energy photons. This is called infrared problem or soft photon problem. If a single photon has a positive mass ν, the energy of a single photon with momentum k is $\sqrt{|k|^2 + \nu^2} \geq \nu$ and hence we can not add energy strictly less than ν. Consequently infrared problem does not occur for massive cases.

The perturbation of discrete eigenvalues has been studied so far, which is called regular perturbation theory, and under some conditions discrete eigenvalues remain to be discrete after adding perturbations. For the Pauli–Fierz Hamiltonian it can be proven that, after adding perturbation, embedded eigenvalues disappear from the real line and turn to be resonances locating on the second Riemann sheet except for the lowest eigenvalue. See Fig. 1.2.

Resonances can be regarded as eigenvalues of the complex dilation $H_{PF}(\theta)$ of H_{PF}. This is established in Bach, Fröhlich and Sigal [7–9]. It can be said that behaviors of embedded eigenvalues are not simple, and excited states are unstable but the ground state is stable.

The purpose of Chap. 3 is to prove the stability of the lowest eigenvalue of H_{PF} for arbitrary values of coupling constant α, where we introduce and review results

[2]The mass of photon is zero.

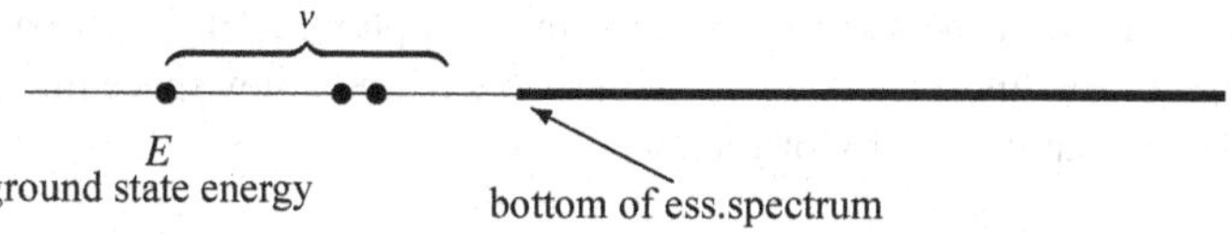

Fig. 1.3 Spectrum of $H_{\rm PF}$ with positive mass $\nu > 0$

obtained by Bach, Fröhlich and Sigal [9] and Griesemer, Lieb and Loss [21]. Introducing an artificial photon mass $\nu > 0$ and the binding condition, it is shown that there exists a positive gap ($\geq \nu$) between the bottom of the spectrum and that of the essential spectrum. See Fig. 1.3. This shows the existence of the ground state Ψ_ν of the massive Pauli–Fierz Hamiltonian for each positive mass $\nu > 0$.

We consider the case of $\nu = 0$. By using a regularized pull-through formula we can estimate the expectation of the number of photons of Ψ_ν by

$$\frac{(\Psi_\nu, N\Psi_\nu)}{\|\langle x\rangle\Psi_\nu\|^2} \leq C_1 \int_{\mathbb{R}^3} \frac{|\hat\varphi(k)|^2}{\sqrt{|k|^2+\nu^2}} dk,$$

where operator N denotes the number operator which counts the number of photons, $\langle x\rangle = \sqrt{|x|^2+1}$, C_1 is a constant independent of ν, and $\hat\varphi$ a cutoff function imposed on A. Furthermore a spatial localization is proven:

$$\frac{\|e^{\delta|x|}\Psi_\nu\|^2}{\|\Psi_\nu\|^2} \leq C_2 \tag{1.3}$$

with some $\delta > 0$ and $C_2 > 0$. We have

$$\frac{(\Psi_\nu, N\Psi_\nu)}{\|\Psi_\nu\|^2} \leq C_1 C_2 \int_{\mathbb{R}^3} \frac{|\hat\varphi(k)|^2}{\sqrt{|k|^2+\nu^2}} dk,$$

and $\hat\varphi(0) = \int_{\mathbb{R}^3} \varphi(x)dx/(2\pi)^{3/2}$ physically describes the total charge divided by $(2\pi)^{3/2}$. In particular $\hat\varphi(0) \neq 0$ should hold. Let $\|\Psi_\nu\| = 1$. We have

$$\sup_{0<\nu}(\Psi_\nu, N\Psi_\nu) \leq C_1 C_2 \int_{\mathbb{R}^3} \frac{|\hat\varphi(k)|^2}{|k|} dk < \infty. \tag{1.4}$$

As far as we know (1.4) was proven by Bach, Fröhlich and Sigal [9] (Theorem 3.23). If

$$\lim_{\nu\downarrow 0}(\Psi_\nu, N\Psi_\nu) = \infty,$$

we say that Ψ_ν has infrared divergence as $\nu \to 0$. (1.4) implies that there is no infrared divergence of Ψ_ν as $\nu \to 0$, and it is said that Ψ_ν has a few "little energy photons". By (1.4) *formally* we can assume that there exists M such that $\Psi_\nu \in \oplus_{n=0}^M L^2(\mathbb{R}^3_x \times \mathbb{R}^{3n}_k)$

for all $0 < \nu$. I.e., we can assume that the number of photons of Ψ_ν is at most M for arbitrary $\nu > 0$. By this and a compact embedding of a Sobolev space into L^2-space, it can be shown that the family of ground states

$$\{\Psi_\nu\}_{\nu>0}$$

is pre-compact in the strong topology. Taking a sequence ν_n such that $\nu_n \to 0$ as $n \to \infty$, we see that Ψ_{ν_n} strongly converges to a non-zero vector $\Psi_{\rm g}$ as $n \to \infty$. We can see that $\Psi_{\rm g}$ is a ground state of $H_{\rm PF}$, and $(\Psi_{\rm g}, N\Psi_{\rm g}) < \infty$.

1.3 The Nelson Model

The Nelson model [49] describes a linear interaction between non-relativistic spinless nucleons and scalar mesons:

$$H_{\rm N} = H_{\rm p} \otimes \mathbb{1} + \mathbb{1} \otimes H_{\rm f} + \phi,$$

where $H_{\rm p}$ is a Schrödinger operator acting[3] in $L^2(\mathbb{R}^3)$ and which describes non-relativistic spinless nucleons, $H_{\rm f}$ the free field Hamiltonian of scalar mesons, and $\phi = \phi_{\hat{\varphi}}$ an interaction with ultraviolet cutoff $\hat{\varphi}$. In Chap. 4 we discuss the existence of the ground state of the Nelson Hamiltonian with/without cutoffs.

Firstly we discuss the Nelson Hamiltonian with cutoffs. Similarly to the Pauli–Fierz Hamiltonian eigenvalues of

$$H_{\rm N,0} = H_{\rm p} \otimes \mathbb{1} + \mathbb{1} \otimes H_{\rm f}$$

are embedded in the continuous spectrum, and the existence of the ground state can be proven in a similar way to the Pauli–Fierz Hamiltonian. We however introduce an alternative method obtained by Gérard [17] in Chap. 4. Independently Spohn [54] also proved the existence of the ground stare of the Nelson Hamiltonian by a functional integration.

By introducing an infrared cutoff parameter $\sigma > 0$ we define H_σ by $H_{\rm N}$ with $\hat{\varphi}$ replaced by $\hat{\varphi}\mathbb{1}_{|k|\geq\sigma}$. It is shown that H_σ is unitary equivalent to some operator $\tilde{H}_\sigma$ plus some well known operator. See Theorem 4.8 for the detail. To prove the existence of the ground state Ψ_σ of H_σ is reduced to prove the existence of the ground state of $\tilde{H}_\sigma$. It can be shown that operator $\chi(\tilde{H}_\sigma)$ is compact, where $\chi \in C_0^\infty((-\infty, \tilde{E}_\sigma + \sigma/2])$ and $\tilde{E}_\sigma = \inf {\rm sp}(\tilde{H}_\sigma)$. Hence $\tilde{H}_\sigma$ has the ground state, and H_σ also has the ground state.

[3] In Chap. 4 the Nelson Hamiltonian is defined on d-dimensional space.

Consider the case of $\sigma = 0$. By a pull-through formula we have

$$\frac{(\Psi_\sigma, N\Psi_\sigma)}{\|\Psi_\sigma\|^2} \le C \int_{|k|>\sigma} \frac{|\hat{\varphi}(k)|^2}{|k|^3} dk. \tag{1.5}$$

The power of the denominator of the integrand on the right-hand side of (1.5) is different from that of (1.4). (1.5) is much singular than (1.4) at $k = 0$. Suppose that $\hat{\varphi}(0) \neq 0$ and $\hat{\varphi}$ is continuous. The integral on the right-hand side above diverges as $\sigma \downarrow 0$ by the singularity at $k = 0$. One can actually show that $(\Psi_\sigma, N\Psi_\sigma) \to \infty$ as $\sigma \downarrow 0$. We can say that Ψ_σ has infrared divergence as $\sigma \to 0$. On the contrary it can be shown in [13, 27] that H_{N} has no ground state if

$$\int_{\mathbb{R}^3} \frac{|\hat{\varphi}(k)|^2}{|k|^3} dk = \infty. \tag{1.6}$$

Equation (1.6) is called infrared singular condition. We introduce the so-called infrared regular condition:

$$\int_{\mathbb{R}^3} \frac{|\hat{\varphi}(k)|^2}{|k|^3} dk < \infty.$$

Let $\|\Psi_\sigma\| = 1$. Under the infrared regular condition we have

$$\sup_{0<\sigma} (\Psi_\sigma, N\Psi_\sigma) \le C \int_{\mathbb{R}^3} \frac{|\hat{\varphi}(k)|^2}{|k|^3} dk < \infty. \tag{1.7}$$

By (1.7) we can construct a certain compact operator T such that

$$\{T\Psi_\sigma\}_{\sigma>0}$$

strongly converges to a non-zero vector $T\Psi_{\mathrm{g}}$ as $\sigma \to 0$. We can see that Ψ_{g} is a ground state of H_{N}.

Secondly we discuss the Nelson Hamiltonian *without* cutoffs. Nelson [49] succeeded in proving the existence of a renormalized Hamiltonian by subtracting an infinite renormalization term. Taking Gross transform of the renormalized Nelson Hamiltonian, Hirokawa, Hiroshima and Spohn [29] show that it also admits a ground state. For positive infrared cutoff $\kappa > 0$ and finite ultraviolet cutoff $\lambda < \infty$, we set

$$\hat{\varphi}(k) = \begin{cases} 0, & |k| < \kappa, \\ (2\pi)^{-3/2}, & \kappa \le |k| \le \lambda, \\ 0, & |k| > \lambda. \end{cases}$$

We introduce the Gross transformation e^T for $\kappa > 0$, which is a unitary operator and $H_{\kappa\lambda}$ is defined by

$$H_{\kappa\lambda} = e^T H_{\mathrm{N}} e^{-T} - R.$$

Here R is a renormalization term which goes to $-\infty$ as $\lambda \to \infty$. It is remarkable that Gross transformed Nelson Hamiltonian $H_{\kappa\lambda}$ is unitary equivalent to $H_{\rm N} - R$ for $\kappa > 0$ but not equivalent for $\kappa = 0$. $H_{\kappa\lambda}$ has the ground state $\Psi_{\kappa\lambda}$ for each $0 < \kappa < \lambda < \infty$. We are interested in studying the spectrum of $H_{0\infty}$. Hirokawa, Hiroshima and Spohn [29] (Theorem 4.33) proved that

$$\frac{(\Psi_{\kappa\lambda}, N\Psi_{\kappa\lambda})}{\|\langle x\rangle\Psi_{\kappa\lambda}\|^2} \leq C_1 \left\{ \int_{|k|\geq 1} \frac{|\hat{\varphi}(k)|^2}{|k|^4} dk + \int_{|k|<1} \left(|k| + \frac{1}{|k|}\right) |\hat{\varphi}(k)|^2 dk \right\}.$$

In a similar way to (1.3) we can show the spatial localization:

$$\frac{\|e^{\delta|x|}\Psi_{\kappa\lambda}\|^2}{\|\Psi_{\kappa\lambda}\|^2} < C_2,$$

where C_2 is independent of both κ and λ. We have

$$\frac{(\Psi_{\kappa\lambda}, N\Psi_{\kappa\lambda})}{\|\Psi_{\kappa\lambda}\|^2} \leq C_1 C_2 \left\{ \int_{|k|\geq 1} \frac{|\hat{\varphi}(k)|^2}{|k|^4} dk + \int_{|k|<1} \left(|k| + \frac{1}{|k|}\right) |\hat{\varphi}(k)|^2 dk \right\}. \tag{1.8}$$

Consider the case of $\kappa = 0$ and $\lambda = \infty$. In this case $\hat{\varphi} = (2\pi)^{-3/2}$, but the right-hand side of (1.8) is finite. Hence we can see that

$$\sup_{0<\kappa<\lambda<\infty} (\Psi_{\kappa\lambda}, N\Psi_{\kappa\lambda}) \leq \frac{C_1 C_2}{(2\pi)^3} \int_{\mathbb{R}^3} \left\{ \frac{1}{|k|^4} \mathbb{1}_{|k|\geq 1} + \left(|k| + \frac{1}{|k|}\right) \mathbb{1}_{|k|<1} \right\} dk < \infty \tag{1.9}$$

and $\Psi_{\kappa\lambda}$ has no infrared divergence as $\kappa \downarrow 0$ and $\lambda \uparrow \infty$. This is the key ingredient to show the existence of a ground state of $H_{0\infty}$. By (1.9), similar to $H_{\rm PF}$ and $H_{\rm N}$, it can be proven that the family of ground states

$$\{\Psi_{\kappa\lambda}\}_{0<\kappa<\lambda<\infty}$$

is pre-compact in the strong topology. Taking subsequences of κ and λ a non-zero strong limits of $\Psi_{\kappa'\lambda'}$ as $\kappa' \to 0$ and $\lambda' \to \infty$ exist, and it is a ground state of $H_{0,\infty}$.

1.4 Spin-Boson Model

Spin-boson model describes a two-level atom interacting with a scalar quantum field. The Hamiltonian is defined by

$$H_{\rm SB} = \varepsilon\sigma_z \otimes \mathbb{1} + \mathbb{1} \otimes H_{\rm f} + \sigma_x \otimes \phi_{\hat{\varphi}},$$

Fig. 1.4 Spectrum of $H_{\mathrm{SB},0}$

where σ_x and σ_z are 2×2 Pauli matrices and $\varepsilon > 0$ denotes a positive constant. H_{SB} can be represented as

$$H_{\mathrm{SB}} = \begin{pmatrix} \varepsilon + H_{\mathrm{f}} & \phi_{\hat{\varphi}} \\ \phi_{\hat{\varphi}} & -\varepsilon + H_{\mathrm{f}} \end{pmatrix}.$$

In Chap. 5 we discuss the existence of the ground state of spin-boson Hamiltonian. $\mathrm{sp}(\varepsilon\sigma_z) = \{-\varepsilon, \varepsilon\}$ means that the energy gap of the two-level atom is $2\varepsilon > 0$. So the spectrum of

$$H_{\mathrm{SB},0} = \varepsilon\sigma_z \otimes \mathbb{1} + \mathbb{1} \otimes H_{\mathrm{f}}$$

is $\mathrm{sp}(H_{\mathrm{SB},0}) = [-\varepsilon, \infty)$ and ε is the unique embedded eigenvalue. See Fig. 1.4.

The existence of the ground state was proven by Arai and Hirokawa [5], where generalized spin-boson models were investigated. It can be also shown by methods of [17, 21], but in Chap. 5 we prove it by an application of Feynman–Kac type formula of semigroup generated by H_{SB}. This method is initiated by Spohn [54], where the Nelson model is investigated. We introduce results obtained in Hirokawa, Hiroshima and Lőrinczi [28] in Chap. 5.

Let $\Omega_{\mathrm{SB}} = \begin{pmatrix} 1 \\ 1 \end{pmatrix} \otimes \Omega$ and $E = \inf \mathrm{sp}(H_{\mathrm{SB}})$, where Ω is the Fock vacuum. Write

$$\Psi_{\mathrm{g}}^T = e^{-T(H_{\mathrm{SB}}-E)}\Omega_{\mathrm{SB}}, \quad T \geq 0,$$

and define $\gamma(T)$ by

$$\gamma(T) = (\Omega_{\mathrm{SB}}, \frac{\Psi_{\mathrm{g}}^T}{\|\Psi_{\mathrm{g}}^T\|})^2.$$

We keep in mind that Ψ_{g}^T is an approximation of the ground state. Formally Ψ_{g}^T converges to the ground state Ψ_{g} as $T \to \infty$ if Ψ_{g} exists. Then we have $\gamma(T) \to (\Omega_{\mathrm{SB}}, \Psi_{\mathrm{g}})^2 > 0$. Conversely it is known that $\lim_{T\to\infty} \gamma(T) > 0$ implies the existence of the ground state. By a path measure we have an expression:

$$\gamma(T) = \frac{\left(\sum_{\sigma\in\mathbb{Z}_2} \mathbb{E}\left[\exp\left(\frac{1}{2}\int_0^T ds \int_0^T W(\sigma_{\varepsilon s}\sigma_{\varepsilon t}, s-t)dt\right)\right]\right)^2}{\sum_{\sigma\in\mathbb{Z}_2} \mathbb{E}\left[\exp\left(\frac{1}{2}\int_{-T}^T ds \int_{-T}^T W(\sigma_{\varepsilon s}\sigma_{\varepsilon t}, s-t)dt\right)\right]}.$$

Here $W = W(X, t)$ is some function, $\mathbb{Z}_2 = \{-1, 1\}$, $\sigma_{\varepsilon t} = \sigma(-1)^{N_{\varepsilon t}}$, $\sigma \in \mathbb{Z}_2$, with a Poisson process $(N_t)_{t\geq 0}$ and $\mathbb{E}$ denotes the expectation with respect to the Poisson process. Since Poisson process takes values in non-negative integers, $\sigma_{\varepsilon t} \in \mathbb{Z}_2$. By using this it can be shown

$$\lim_{T\to\infty} \gamma(T) > \exp\left(-\frac{1}{2}\int_{\mathbb{R}^d} \frac{|\hat{\varphi}(k)|^2}{|k|^3} dk\right). \tag{1.10}$$

This implies that $\lim_{T\to\infty} \gamma(T) > 0$ if the infrared regular condition is assumed.

1.5 Infrared Problem

As is seen in Sects. 1.2, 1.3 and 1.4, it is crucial to show

$$\frac{(\Psi_\xi, N\Psi_\xi)}{\|\Psi_\xi\|^2} < C$$

uniformly in parameter ξ in order to prove the existence of ground states. Here $\xi = \nu, \sigma, \{\kappa, \lambda\}$. This accords with a physical folklore, i.e., if there are a few soft-photons (soft-bosons), there exists the ground state.

There is no infrared divergence for the Pauli–Fierz Hamiltonian $H_{\rm PF}$ ($\nu = 0$) and the Gross transformed Nelson Hamiltonian $H_{\kappa\lambda}$ with $0 \le \kappa < \lambda \le \infty$, and one can show the existence of ground states of both Hamiltonians without the infrared regular condition. See (1.4) and (1.8). It is emphasised that $H_{\rm N}$ and $H_{\kappa\lambda}$ are unitary equivalent if $\kappa > 0$.

On the other hand there is infrared divergence for the Nelson Hamiltonian $H_{\rm N}$ with $\sigma = 0$. See (1.5). We need infrared regular condition to show the existence of the ground state for $H_{\rm N}$ with $\sigma = 0$. If infrared singular condition is satisfied, $H_{\rm N}$ has no ground state.

Although both $H_{\rm N}$ and $H_{\rm SB}$ look in the same, when $\sigma = 0$, it is proven in [24] that $H_{\rm SB}$ has no infrared divergence and it admits the ground state. This is different form $H_{\rm N}$ with $\sigma = 0$. See Table 1.1.

Table 1.1 Existence of ground state for $\int_{\mathbb{R}^d} \frac{|\hat{\varphi}(k)|^2}{|k|^3} dk = I \le \infty$

	$I < \infty$	$I = \infty$
$H_{\rm PF}$	Exist	Exist
$H_{\rm N}$	Exist	Not exist
$H_{0\infty}$	Exist	Exist
$H_{\rm SB}$	Exist	Exist

1.6 Enhanced Binding

It is expected that an interaction with quantum field enhances binding and sufficiently strong coupling consequently produces a ground state. This phenomena are called enhanced binding. Chap. 6 is devoted to showing the enhanced binding for the Pauli–Fierz model and the Nelson model. We review papers Hiroshima and Spohn [38], Hiroshima, Suzuki and Spohn [39] and Hiroshima and Sasaki [37]. A mathematical setting of problem is as follows. Suppose that zero-coupling Hamiltonian $H_{0,\sharp}$ has no ground state for $\sharp = \mathrm{PF}, \mathrm{N}$. Does the full Hamiltonian $H_\sharp$ admits a ground state for sufficiently strong couplings?

It is remarkable that mechanisms of enhanced binding for H_{PF} and H_{N} are different from each others. Enhanced binding is caused by the effective mass for H_{PF} but effective potential for H_{N}. Hence enhanced binding for H_{PF} occurs for a single particle, but that for H_{N} occurs for more than two particles.

1.6.1 Enhanced Binding for the Pauli–Fierz Hamiltonian

For enhanced binding for the Pauli–Fierz Hamiltonian, instead of H_{PF}, we show the enhanced binding for $H_{\mathrm{PF}}^{\mathrm{dip}}$ for simplicity. Here $H_{\mathrm{PF}}^{\mathrm{dip}}$ denotes the Pauli–Fierz Hamiltonian with dipole approximation. Formally we can see

$$H_{\mathrm{PF}}^{\mathrm{dip}} \sim \left(-\frac{1}{2m_{\mathrm{eff}}}\Delta + V\right) \otimes \mathbb{1} + \mathbb{1} \otimes H_{\mathrm{rad}} + R_{\mathrm{PF}}, \tag{1.11}$$

where

$$m_{\mathrm{eff}} = m_{\mathrm{eff}}(\alpha^2) = m + \frac{2}{3}\alpha^2 \|\hat{\varphi}/\omega\|^2$$

is an effective mass and R_{PF} a remainder term. A rigorous derivation of (1.11) is found in [2, 38]. From the right-hand side of (1.11) we can read a following physical phenomena: as the particle binds photons it acquires an effective mass $m_{\mathrm{eff}} = m_{\mathrm{eff}}(\alpha^2)$ which is increasing in $|\alpha|$. Suppose that $-\frac{1}{2m}\Delta + V$ has no ground state. From (1.11) we expect that $H_{\mathrm{PF}}^{\mathrm{dip}}$ has no ground state for small $|\alpha|$ but has a ground state for large $|\alpha|$. See Fig. 1.5.

To show the enhanced binding for $H_{\mathrm{PF}}^{\mathrm{dip}}$ we take momentum lattice approximation. This method is used in Glimm and Jaffe [19] and applied to related models in [9] and [5]. The boson Fock space $\mathscr{F}(\ell^2(\mathbb{Z}^3/a^3))$ over the lattice $\ell^2(\mathbb{Z}^3/a^3)$ with spacing $1/a$ can be regarded as the subspace of $\mathscr{F}(L^2(\mathbb{R}^3))$. The procedure is (1)–(3) below:

(1) Lattice approximated Hamiltonian $H_{a,L}$ satisfying $\lim_{a\to\infty, L\to\infty} H_{a,L} = H_{\mathrm{PF}}^{\mathrm{dip}}$ is defined on $\mathscr{F}(\ell^2(\mathbb{Z}^3/a^3))$, where L denotes the field momentum cutoff.
(2) Enhanced biding for $H_{a,L}$ is proven.

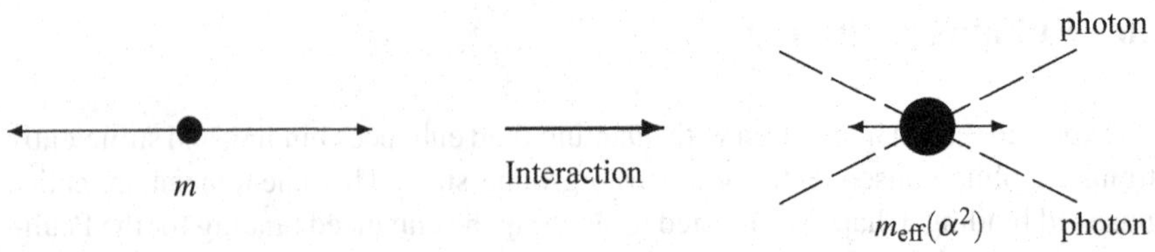

Fig. 1.5 Effective mass enhances binding for $H_{\rm PF}^{\rm dip}$

(3) Enhanced biding for $H_{\rm PF}^{\rm dip}$ is also shown by a limiting argument.

On the other hand, the absence of ground state for small $|\alpha|$ is proven by estimating a Birman-Schwinger kernel.

1.6.2 Enhanced Binding for the Nelson Hamiltonian

For enhanced binding for the Nelson Hamiltonian, we consider N-body Nelson Hamiltonian which describes N-particles linearly coupled to a scalar field:

$$H_{\rm N} = \left(-\sum_{j=1}^{N} \frac{1}{2m_j}\Delta_j + V\right) \otimes \mathbb{1} + \mathbb{1} \otimes H_{\rm f} + \sum_{j=1}^{N} g_j \phi_{\hat{\varphi}_j}(x_j).$$

The jth particle has mass m_j, $j = 1, \ldots, N$, and any pair of particles does *not* interact via external potential, i.e.,

$$V(x_1, \ldots x_N) = \sum_{j=1}^{N} V_j(x_j).$$

We can however formally see that

$$H_{\rm N} \sim \left(\sum_{j=1}^{N}\left(-\frac{1}{2m_j}\Delta_j + V_j\right) + V_{\rm eff}\right) \otimes \mathbb{1} + \mathbb{1} \otimes H_{\rm f} + R_{\rm N}.$$

Here

$$V_{\rm eff}(x) = -\frac{1}{4}\sum_{i \neq j}^{N} g_i g_j \int_{\mathbb{R}^d} \frac{\hat{\varphi}_i(-k)\hat{\varphi}_j(k)}{\omega(k)} e^{-ik\cdot(x_i - x_j)} dk$$

is an attractive effective potential which connects any pair of particles, and $R_{\rm N}$ is a remainder term. If N particles attractively interact with each other by an effective potential provided by the scalar quantum field, particles close up with each other and

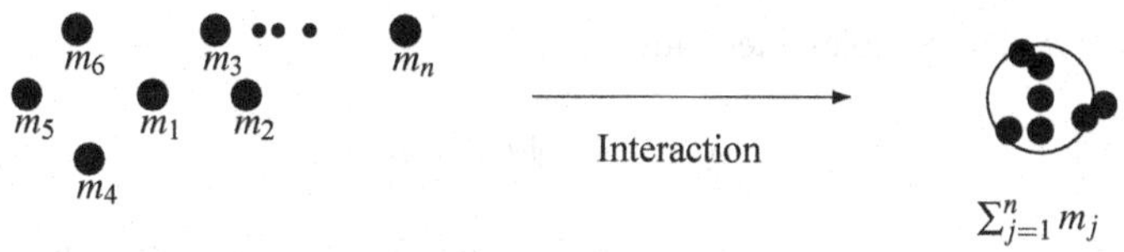

Fig. 1.6 Effective potential enhances binding for H_{N}

seem to behave like as a single particle with mass $\sum_{j=1}^{N} m_j$ and H_{N} has the ground state. See Fig. 1.6.

Enhanced binding for the N-body Nelson Hamiltonian H_{N} is shown by the hybrid of methods [17, 21].

1.7 Compactness

As was seen in the Pauli–Fierz Hamiltonian and the Nelson Hamiltonian, in order to show the existence of the ground state Ψ of H, we firstly show the existence of ground state Ψ_ξ with parameter $\xi > 0$, where $\xi = \nu$ for H_{PF} and $\xi = \sigma$ for H_{N}, etc.

$$H_\xi \Psi_\xi = E_\xi \Psi_\xi.$$

Since Ψ_ξ is normalized, $\|\Psi_\xi\| = 1$, $\{\Psi_\xi\}_{\xi>0}$ includes a *weakly* convergent sequence $(\Psi_{\xi_k})_{k=1}^\infty$. Let

$$w - \lim_{k\to\infty} \Psi_{\xi_k} = \Psi.$$

By Proposition 3.34 we know that Ψ is a ground state of H if $\Psi \neq 0$. To see $\Psi \neq 0$ we have two strategies.

(1) We show that $\{\Psi_\xi\}_{\xi>0}$ is a pre-compact set.
(2) We find a compact operator T such that $s - \lim_{k\to\infty} T\Psi_{\xi_k} \neq 0$.

Chapter 2 is devoted to introducing useful tools on compactness. Ascoli–Arzela theorem gives an example of a compact set in $C(K)$.[4] This theorem can be extended to L^p-version, which is known as Kolmogorov–Riesz–Fréchet theorem. This theorem can be applicable to see that $\{\Psi_\xi\}_{\xi>0}$ is pre-compact in [36]. Technically the most useful compact operators are of the form $Q(x)P(-i\nabla_x)$ with some function P and Q decaying to zero. In [17] the compactness of operators

$$\Gamma(Q(k))\Gamma(P(-i\nabla_k)), \quad \Gamma(P(-i\nabla))(H_{\mathrm{f}} + \mathbb{1})^{-1}$$

on any finite particle space is used. See details in Chap. 2. Another important compact operator is an embedding. Let d be the space dimension and $U \subset \mathbb{R}^d$ bounded set.

[4]The set of continuous functions on a compact set K.

Let $1 \leq p < d$. The Sobolev inequality

$$\|f\|_{L^q(U)} \leq C\|f\|_{W^{1,p}(U)}$$

is satisfied for $q = dp/(d-p)$, which means that $W^{1,p}(U)$ is continuously embedded in $L^q(U)$. Weakening topology of $L^q(U)$, we can see that $W^{1,p}(U)$ is compactly embedded in $L^q(U)$ for $q < dp/(d-p)$, i.e., identity

$$\iota : W^{1,p}(U) \to L^q(U)$$

is compact. This is known as Rellich–Kondrachov compactness theorem. Thus from bounded set in $W^{1,p}(U)$ we can extract a strong convergent subsequence in $L^q(U)$, which trick will be used to show the existence of the ground state in [21].

Chapter 2
Preliminaries

Abstract In this chapter we introduce fundamental tools used throughout this book. Compact operators on Banach spaces and compact embeddings of Sobolev spaces of the form $W^{1,p}(U) \subset\subset L^q(U)$ are reviewed, which can be applied to study perturbations of eigenvalues embedded in the continuous spectrum of selfadjoint operators which describe Hamiltonians in quantum field theory. The boson Fock space $\mathscr{F}(\mathscr{W})$ over Hilbert space $\mathscr{W}$ is defined. Creation operators $a(f)$, annihilation operators $a^\dagger(f)$, second quantization $\Gamma(T)$ and differential second quantization $\mathrm{d}\Gamma(h)$ are introduced as operators in $\mathscr{F}(\mathscr{W})$. We also define operator $\mathrm{d}\Gamma(k,h)$ being an extension of $\mathrm{d}\Gamma(h)$ and discuss localizations in $\mathscr{F}(\mathscr{W})$ via the canonical identification $\mathscr{F}(\mathscr{W}_1 \oplus \mathscr{W}_2) \cong \mathscr{F}(\mathscr{W}_1) \otimes \mathscr{F}(\mathscr{W}_2)$. Finally we review compact operators of the form $Q(x)P(-i\nabla)$ in $L^2(\mathbb{R}^d)$ and $(\mathrm{d}\Gamma(|k|) + \mathbb{1})^{-1}\Gamma(F(-i\nabla_k))$ in $\mathscr{F}(L^2(\mathbb{R}^d))$, and demonstrate their applications.

2.1 Compact Operators

2.1.1 Compact Operators Basics

Let $\mathscr{H}$ be a Hilbert space. The scalar product on $\mathscr{H}$ is denoted by $(f, g)_{\mathscr{H}}$ and the norm by $\|f\|_{\mathscr{H}} = \sqrt{(f, f)_{\mathscr{H}}}$. Here and in what follows it is assumed that the scalar product is anti-linear in the first component and linear in the second one. Similarly the norm defined on Banach space X is denoted by $\|\cdot\|_X$. In what follows we omit subscript K of the scalar product $(\cdot,\cdot)_K$ and the norm $\|\cdot\|_K$ unless confusion may arise.

Let X and Y be Banach spaces, and $B(X, Y)$ denotes the set of bounded operators from X to Y. We write $B(X)$ for $B(X, X)$. $B(X, Y)$ is also a Banach space under the operator norm. There is a class in $B(X, Y)$, which is called compact operators or completely continuous operators.

Definition 2.1 (*Compact operator*) $T \in B(X, Y)$ is a compact operator if and only if T maps a bounded set $\{f_n\}_{n=1}^\infty \subset X$ to a pre-compact set $\{Tf_n\}_{n=1}^\infty \subset Y$, i.e., $\{Tf_n\}_{n=1}^\infty$

F. Hiroshima, *Ground States of Quantum Field Models*,
SpringerBriefs in Mathematical Physics,
https://doi.org/10.1007/978-981-32-9305-2_2

contains a strongly convergent subsequence. The set of compact operators is denoted by $B_c(X, Y)$, and we set $B_c(X, X) = B_c(X)$.

Although historically compact operators have been developed to solve integral equations known as Fredholm theory appeared in 1903 [16], compact operators have played an important role in studying a perturbation of embedded eigenvalues of selfadjoint operators appearing in quantum theory. We list algebraic properties of compact operators.

(1) $B_c(X, Y)$ is a closed subspace of $B(X, Y)$. In particular it is also a Banach space with the same norm as that of $B(X, Y)$. Hence the operator norm convergence of a sequence of compact operators is also compact.
(2) The product of a compact operator and a bounded operator is also compact. More precisely let $T \in B_c(X, Y)$, $A \in B(Y, Z)$ and $B \in B(W, X)$, then $AT \in B_c(X, Z)$ and $TB \in B_c(W, Y)$. In particular $B_c(X)$ is a two-sided ideal of $B(X)$, and $B(X)/B_c(X)$ is called Calkin algebra.
(3) $T \in B(X, Y)$ implies $T^* \in B_c(Y^*, X^*)$. Hence if X is a Hilbert space, then $T \in B(X)$ if and only if $T^* \in B(X)$ under the identification $X \cong X^*$.

Example 2.1 (1) If either X or Y is a finite dimensional space, then $T \in B(X, Y)$ is a compact operator.
(2) Let $P \in B(X)$ be a projection. P is compact if and only if $\mathrm{Ran}\,P$ is a finite dimensional space. Then the identity operator $\mathbb{1}_X \in B(X)$ is in general not compact, but when X is a finite dimensional space, $\mathbb{1}_X$ turns to be a compact operator.
(3) A finite rank operator[1] is compact. If $\{A_n\}_{n=1}^{\infty}$ is a sequence of finite rank operators and $A = \lim_{n\to\infty} A_n$ in the operator norm, A is also compact.
(4) Let E and F be Lebesgue measurable subsets of $\mathbb{R}$. $L^1(E)$ denotes the set of integrable functions on E and $C(F)$ the set of continuous functions on F. We define the linear operator T by

$$Tf(x) = \int_E k(y, x) f(y) dy, \quad x \in F$$

with an integral kernel k. Suppose that E and F are compact sets in $\mathbb{R}$ and k is continuous on $\mathbb{R} \times \mathbb{R}$. Then $T \in B(L^1(E), C(F))$, and by Ascoli–Arzela theorem (Theorem 2.12) $\{Tf_n\}_{n=1}^{\infty}$ contains a convergent sequence if $\{f_n\}_{n=1}^{\infty} \subset L^1(E)$ is bounded. Thus T is compact.
(5) $A \in B_c(X)$ and $B \in B_c(Y)$ imply that $A \otimes B \in B_c(X \otimes Y)$.

An important property of compact operators is to map a weakly convergent sequence into a strongly convergent sequence.

Proposition 2.2 *(1) Suppose that $\{f_n\}_{n=1}^{\infty} \subset X$ weakly converges to f as $n \to \infty$ and $T \in B_c(X, Y)$. Then $\{Tf_n\}_{n=1}^{\infty}$ strongly converges to Tf in Y. (2) Suppose*

[1] T is a finite rank operator if and only if $\mathrm{Ran}\,T$ is a finite dimensional space.

that X is reflexive. Assume that $T \in B(X, Y)$ satisfies that for any weakly convergent sequence $\{f_n\}_{n=1}^\infty \subset X$, $\{Tf_n\}_{n=1}^\infty$ is a strongly convergent sequence. Then $T \in B_c(X, Y)$.

Proof (1) By the uniform boundedness theorem $\sup_{n\in\mathbb{N}} \|f_n\|_X < \infty$. Let $Tf_n = g_n$ and $F \in X^*$. Then $F(g_n) - F(g) = (T^*F)(f_n - f) \to 0$ as $n \to \infty$, and hence $\{g_n\}_{n=1}^\infty$ weakly converges to g. Suppose that $\{g_n\}_{n=1}^\infty$ does not strongly converge to g in Y. There exists $\delta > 0$ and a subsequence $\{g_{n_k}\}_{k=1}^\infty$ of $\{g_n\}_{n=1}^\infty$ such that $\|g_{n_k} - g\| > \delta$ for any n_k. Since $\{f_{n_k}\}_{k=1}^\infty$ is a bounded sequence and T is compact, $\{g_{n_k}\}_{k=1}^\infty$ has a convergent subsequence which converges to h, but h can not be g because of $\|h - g\| \geq \delta$. The subsequence however weakly converges to h. Hence $h = g$. This is a contradiction. Then $\{g_n\}_{n=1}^\infty$ strongly converges to g.

(2) Let $\{g_n\}_{n=1}^\infty \subset X$ be such that $\sup_{n\in\mathbb{N}} \|g_n\| < \infty$. Since X is reflexive, there exists a subsequence $\{g_{n_k}\}_{k=1}^\infty$ such that $g_{n_k} \to g$ weakly as $k \to \infty$. Thus $Tg_{n_k} \to Tg$ strongly as $k \to \infty$, which implies that T is compact. ■

2.1.2 Spectrum of Compact Operators

The spectrum of a compact operator has a simple structure which is an analogue of that of an operator on a finite dimensional space, say matrix. Let $\mathrm{sp}(K)$ (resp. $\mathrm{sp_p}(K)$, $\mathrm{sp_{disc}}(K)$, $\mathrm{sp_{ess}}(K)$) be the spectrum (resp. point spectrum, discrete spectrum, essential spectrum) of closable operator K. The following fact is known as Riesz-Schauder theorem:

Theorem 2.3 (Riesz-Schauder theorem) *For $T \in B_c(X)$, $\mathrm{sp}(T) \setminus \{0\}$ is discrete, i.e., $z \in \mathrm{sp}(T) \setminus \{0\}$ is an eigenvalue of T with a finite multiplicity, and 0 is possibly the unique accumulation point of $\mathrm{sp}(T)$.*

Proof See [52]. ■

Now let us consider compact operators on a separable Hilbert space $\mathscr{H}$. Let T be a selfadjoint compact operator on $\mathscr{H}$. By Riesz-Shauder theorem we can see that $\mathscr{H}$ has a complete orthonormal basis $\{\phi_n\}_{n=1}^\infty$ of eigenvectors of T so that $T\phi_n = \lambda_n\phi_n$ and $\lambda_n \to 0$ as $n \to 0$. Hence we have $Tf = \sum_{n=1}^\infty \lambda_n(\phi_n, f)\phi_n$.

Proposition 2.4 (Canonical form for compact operators) *Let $T \in B_c(\mathscr{H})$. Then T has the operator norm convergent expansion*

$$T = \sum_{n=1}^{N} \mu_n(T)(\phi_n, \cdot)\psi_n.$$

Here $N \leq \infty$, $\mu_n(T) > 0$, $n = 1, 2, \ldots$, is a non-increasing sequence and both $\{\phi_n\}_{n=1}^\infty$ and $\{\psi_n\}_{n=1}^\infty$ are orthonormal bases but not necessarily complete. Moreover $\lim_{n\to\infty} \mu_n(T) = 0$ if $N = \infty$.

Proof By the polar decomposition $T = U|T|$ we have $|T| = U^*T$, where U is the isometry on $\mathrm{Ran}|T|$. Since $|T|$ is selfadjoint, $|T| = \sum_{n=1}^{N} \mu_n(T)(\phi_n, \cdot)\phi_n$, where $\mu_n(T)$ is non-zero eigenvalues of $|T|$ and ϕ_n the corresponding eigenvector. Since U is isometry on $\mathrm{Ran}|T|$, $\psi_n = U\phi_n$ is also orthonormal. Then the proposition follows. ■

By the canonical form for compact operators we can see that any compact operator can be approximated by finite rank operators in the operator norm. Let $\mathscr{H}$ be a separable Hilbert space, $T \in B_c(\mathscr{H})$ and $\mu_1(T) \geq \mu_2(T) \geq \ldots$ the sequence of eigenvalues of $|T|$ arranged in non-increasing order with each eigenvalue occurring as often as its multiplicity. Let $1 \leq p < \infty$. We say that $T \in B_p(\mathscr{H})$ if and only if $\left(\sum_{n=1}^{\infty} \mu_n(T)^p\right)^{1/p} < \infty$. We set

$$\|T\|_p = \left(\sum_{n=1}^{\infty} \mu_n(A)^p\right)^{1/p}.$$

In particular $B_1(\mathscr{H})$ is called the trace class, and $B_2(\mathscr{H})$ the Hilbert–Schmidt class. Let $\{e_n\}_{n=1}^{\infty}$ be an arbitrary complete orthonormal basis of $\mathscr{H}$. It is known that T is trace class if and only if $\sum_{n=1}^{\infty} |(e_n, Te_n)| < \infty$ and Hilbert–Schmidt class if and only if T^*T is trace class. See e.g., [52, VI.6].

2.1.3 Compact Operators on $L^2(X)$

Let $\mathscr{H} = L^2(X, d\mu)$ with a separable measure space (X, μ). $\mathscr{H}$ is a Hilbert space with scalar product $(f, g)_{\mathscr{H}} = \int_X \bar{f}(x)g(x)d\mu(x)$. We shall show that a Hilbert–Schmidt operator on $\mathscr{H}$ can be characterized in terms of integral kernels. Suppose that $T \in B_2(\mathscr{H})$. We formally see that $T = \sum_{n=1}^{\infty} \lambda_n(\phi_n, \cdot)\psi_n$ and

$$(g, Tf) = \int_{X\times X} \bar{g}(x)k(x, y)f(y)d\mu(x)d\mu(y),$$

where $k(x, y) = \sum_{n=1}^{\infty} \lambda_n\bar{\phi}_n(y)\psi_n(x)$ and

$$\int_{X\times X} |k(x, y)|^2 d\mu(x)d\mu(y) = \sum_{n=1}^{\infty} \lambda_n^2 < \infty. \tag{2.1}$$

On the other hand suppose (2.1). Then for any complete orthonormal basis $\{e_n\}_{n=1}^{\infty}$ we have

$$\begin{aligned}\sum_{n=1}^{\infty}(e_n, T^*Te_n) &= \int_X d\mu(x) \int_{X\times X} k(x, y)\bar{k}(x, y')\bar{e}_n(y)e_n(y')d\mu(y)d\mu(y') \\ &= \int_X \|k(x, \cdot)\|^2 d\mu(x) = \int_{X\times X} |k(x, y)|^2 d\mu(x)d\mu(y) < \infty.\end{aligned}$$

We conclude that $T \in B_2(\mathscr{H})$. This formal description can be exactly established.

Proposition 2.5 *$T \in B(\mathscr{H})$ is a Hilbert–Schmidt operator if and only if there exists a unique integral kernel $k \in L^2(X \times X, d\mu \otimes d\mu)$ such that*

$$Tf(x) = \int_{X\times X} k(x, y) f(y) d\mu(y).$$

If T is Hilbert–Schmidt, $\|T\|_2 = \|k\|_{L^2(X\times X)}$ follows.

Proof See [52, Theorem VI.23]. ∎

Let us consider the operator of the form $T = f(x)g(-i\nabla)$, where $f, g \in L^2(\mathbb{R}^d)$. To present T as an integral operator we use the Fourier transform.

Schwartz space $\mathscr{S}(\mathbb{R}^d)$ is the set of infinitely differentiable complex-valued function $f(x)$ on $\mathbb{R}^d$ for which $\sup_{x\in\mathbb{R}^d} |x^\alpha D^\beta f(x)| < \infty$ for all $\alpha = (\alpha_1, \ldots, \alpha_d)$ and $\beta = (\beta_1, \ldots, \beta_d)$. Here α_i and β_j are nonnegative integers and $x^\alpha = x_1^{\alpha_1} \cdots x_d^{\alpha_d}$ and $D^\beta = \partial x_1^{\beta_1} \cdots \partial x_d^{\beta_d}$. $\mathscr{S}(\mathbb{R}^d)$ is a dense subspace of $L^2(\mathbb{R}^d)$. The Fourier transformation $F : \mathscr{S}(\mathbb{R}^d) \to \mathscr{S}(\mathbb{R}^d)$ is a bijective linear map defined by

$$Ff(k) = \frac{1}{(2\pi)^{d/2}} \int_{\mathbb{R}^d} f(x) e^{-ikx} dx.$$

The Fourier transform of f, $Ff(k)$, is written as $\hat{f}(k)$. On the other hand the inverse Fourier transform is given by

$$F^{-1} f(x) = \frac{1}{(2\pi)^{d/2}} \int_{\mathbb{R}^d} f(k) e^{+ikx} dk.$$

The inverse Fourier transform of f, $F^{-1}f(x)$, is written as $\check{f}(x)$. They satisfy $FF^{-1} = F^{-1}F = I$ on $\mathscr{S}(\mathbb{R}^d)$ and $(Ff, Fg) = (f, g)$, where (f, g) denotes the scalar product on $L^2(\mathbb{R}^d)$. Since $\mathscr{S}(\mathbb{R}^d)$ is dense in $L^2(\mathbb{R}^d)$, the Fourier transformation F can be extended to the unitary operator $\bar{F}$ on $L^2(\mathbb{R}^d)$. We denote the same notation F instead of $\bar{F}$ in what follows. The dual of $\mathscr{S}(\mathbb{R}^d)$ is denoted by $\mathscr{S}'(\mathbb{R}^d)$ and the Fourier transform F is defined on $\mathscr{S}'(\mathbb{R}^d)$ by $FT(\phi) = T(F^{-1}\phi)$ for $T \in \mathscr{S}'(\mathbb{R}^d)$ and $\phi \in \mathscr{S}(\mathbb{R}^d)$. It can be seen that $F(-i\nabla_j f)(k) = k_j \hat{f}(k)$ for $f \in D(-i\nabla_j)$ and

$$g(-i\nabla)h(x) = (2\pi)^{-d/2} \int_{\mathbb{R}^d} \check{g}(x - y) h(y) dy, \quad g, h \in L^2(\mathbb{R}^d).$$

We have the proposition below.

Proposition 2.6 *Let $f, g \in L^2(\mathbb{R}^d)$ and define $T = f(x)g(-i\nabla)$ on $L^2(\mathbb{R}^d)$. Then T is a Hilbert–Schmidt operator and $\|T\|_2 = (2\pi)^{-d/2}\|f\|_{L^2(\mathbb{R}^d)}\|g\|_{L^2(\mathbb{R}^d)}$.*

Proof By the definition, operator T is the integral operator with the integral kernel explicitly given by $k(x, y) = (2\pi)^{-d/2} f(x)\check{g}(x - y)$ and we can see that $\|k\|_{L^2(\mathbb{R}^d \times \mathbb{R}^d)} = (2\pi)^{-d/2}\|f\|_{L^2(\mathbb{R}^d)}\|g\|_{L^2(\mathbb{R}^d)} < \infty$. Then the proposition follows. ■

By the interpolation we can also show more strong statements.

Proposition 2.7 *We suppose that* $f, g \in L^p(\mathbb{R}^d)$ *for* $2 \leq p < \infty$. *Then we have* $f(x)g(-i\nabla) \in B_p(L^2(\mathbb{R}^d))$ *and* $\|f(x)g(-i\nabla)\|_p \leq (2\pi)^{-d/2}\|f\|_{L^p(\mathbb{R}^d)}\|g\|_{L^p(\mathbb{R}^d)}$.

Proof See [53, Theorem 4.1]. ■

To see the compactness of the form $f(x)g(-i\nabla)$, we do not need integrable conditions for f and g. Let $L^{\infty,0}(\mathbb{R}^d) = \{f \in L^\infty(\mathbb{R}^d) | f(x) \to 0 \text{ as } |x| \to \infty\}$.

Corollary 2.8 *Let* $P, Q \in L^{\infty,0}(\mathbb{R}^d)$. *Then* $Q(x)P(-i\nabla)$ *is a compact operator.*

Proof Let $\chi_\lambda \in C_0^\infty(\mathbb{R}^d)$ with

$$\chi_\lambda(x) = \begin{cases} 1, & |x| < \lambda, \\ 0, & |x| > \lambda + 1. \end{cases}$$

$\chi_\lambda(x)\chi_\lambda(-i\nabla)$ is a compact operator by Proposition 2.6. Since $Q\chi_\lambda \to Q$ and $\chi_\lambda P \to P$ as $\lambda \to \infty$ in $L^\infty(\mathbb{R}^d)$, $Q(x)P(-i\nabla) = \lim_{\lambda\to\infty} Q(x)\chi_\lambda(x)\chi_\lambda(-i\nabla) P(-i\nabla)$ in the operator norm. This yields that $Q(x)P(-i\nabla)$ is compact. ■

Example 2.9 Let $H_\mathrm{p} = -\frac{1}{2}\Delta + V$ be a Schrödinger operator in $L^2(\mathbb{R}^d)$. Suppose that $V \in L^{\infty,0}(\mathbb{R}^d)$. Since $V(-\frac{1}{2}\Delta + \mathbb{1})^{-1}$ is compact by Corollary 2.8, H_p is self-adjoint on $D(-\Delta)$ and V is relatively compact with respect to $-\frac{1}{2}\Delta$. In particular $\mathrm{Sp}_\mathrm{ess}(H_\mathrm{p}) = \mathrm{Sp}_\mathrm{ess}(-\Delta) = [0, \infty)$.

2.1.4 Carleman Operators

The most important class of compact operators is the Hilbert–Schmidt class. We can see that a Hilbert–Schmidt operator on $L^2(X)$ is an integral operator with integral kernel k such that $k \in L^2(X \times X)$. More generally we can consider operators from a Hilbert space $\mathscr{H}$ to $L^2(X)$, which is referred to Carleman operators.

Proposition 2.10 *Let* $T \in B(\mathscr{H}, L^2(X))$, *where* $\mathscr{H}$ *is a separable Hilbert space and* (X, μ) *a separable measure space.* T *is a Hilbert–Schmidt operator if and only if there exists a function* $\kappa : X \to \mathscr{H}$ *such that* $\|\kappa(\cdot)\|_{\mathscr{H}} \in L^2(X)$ *and*

$$Tf(x) = (\kappa(x), f)_{\mathscr{H}} \text{ for a.e. } x \text{ in } X.$$

Proof Let $\{e_n\}_{n=1}^\infty \subset \mathscr{H}$ be an orthonormal basis of $\mathrm{Ker}(T)^\perp$. We have

$$\sum_{n=1}^\infty \int_X |Te_n(x)|^2 d\mu = \sum_{n=1}^\infty \|Te_n\|^2_{L^2(X)} < \infty.$$

By the monotone convergence theorem we see that $\sum_{n=1}^{\infty} |Te_n(x)|^2 < \infty$ for $x \in X \setminus N$ with some null set N, i.e., $\mu(N) = 0$. Define the function $\kappa : X \to \mathscr{H}$ by

$$\kappa(x) = \begin{cases} \sum_{n=1}^{\infty} \overline{Te_n(x)} e_n, & x \in X \setminus N, \\ 0, & x \in N. \end{cases}$$

Hence we have

$$(\kappa(x), f) = \sum_{n=1}^{\infty} (e_n, f) Te_n(x) = Tf(x).$$

Moreover $\|\kappa(x)\|^2 = \sum_{n=1}^{\infty} |Te_n(x)|^2$ yields that $\|\kappa(\cdot)\| \in L^2(X)$.

Conversely we assume that there exists κ such that $Tf(x) = (\kappa(x), f)$ a.e. and $\int_X \|\kappa(x)\|^2 d\mu < \infty$. Precisely $Tf(x) = (\kappa(x), f)$ holds for $x \in X \setminus N_f$ with null set N_f. Hence $\|Tf(x)\| \leq \|f\| \|\kappa(x)\|$ follows for $x \in X \setminus N_f$. Let us define a linear operator $S : \text{L.H.}\{e_1, \ldots, e_n\} \to L^2(X)$ by $S(\sum_{j=1}^n a_j e_j)(x) = \sum_{j=1}^n a_j Te_j(x)$ with $a_j \in \mathbb{C}$. Here L.H. denotes a shorthand for the linear hull. Hence for each $x \in X \setminus M$ with $M = \cup_{j=1}^n N_{e_j}$, $\text{L.H.}\{e_1, \ldots, e_n\} \ni f \mapsto Sf(x) \in \mathbb{C}$ is bounded with $|Sf(x)| \leq \|\kappa(x)\| \|f\|$. Then there exists $S(x) \in \text{L.H.}\{e_1, \ldots, e_n\}$ such that $Sf(x) = (S(x), f)$ and $\|S(x)\| \leq \|\kappa(x)\|$ for $x \in X \setminus M$. It follows that

$$\sum_{j=1}^{n} \|Te_j\|^2 = \sum_{j=1}^{n} \|Se_j\|^2 = \int_X \sum_{j=1}^{n} |(S(x), e_j)|^2 d\mu \leq \int_X \|\kappa(x)\|^2 d\mu < \infty.$$

Then $\sum_{j=1}^{\infty} \|Te_j\|^2 \leq \int_X \|\kappa(x)\|^2 d\mu < \infty$ and the proof is complete. ∎

Definition 2.11 (*Carleman operator*) A linear operator T from a Hilbert space $\mathscr{H}$ to $L^2(X)$ is called a Carleman operator if there exists a function $\kappa : X \to \mathscr{H}$ such that $Tf(x) = (\kappa(x), f)$ a.e.

We refer to e.g., [56, Sect. 6.2] for the detail of Carleman operators. By Proposition 2.10 every Hilbert–Schmidt operator from $\mathscr{H}$ to $L^2(X)$ is a Carleman operator.

2.2 Compact Embeddings

2.2.1 Strong Compactness in L^p Spaces

If $D \subset L^p(\mathbb{R}^d)$ is a bounded set and T a compact operator on it, set TD is pre-compact in the strong topology. We are interested in deciding whether a subset of $L^p(U)$ is pre-compact in $L^p(U)$ in the strong topology. Ascoli–Arzela theorem answers the same question in $C(K)$, the space of continuous functions on a compact metric space (K, d).

Theorem 2.12 (Ascoli–Arzela theorem) *Let K be a compact metric space and D be a bounded subset of $C(K)$. Assume that D is uniformly equicontinuous.*[2] *Then D is pre-compact in $C(K)$, i.e., $\bar{D}$ is compact.*

The following theorem is L^p-version of Ascoli–Arzela theorem.

Theorem 2.13 (Kolmogorov-Riesz-Fréchet theorem) *Suppose that $D \subset L^p(\mathbb{R}^d)$ is bounded and $1 \le p < \infty$. Assume that for $\forall \varepsilon > 0$ there exists $\delta > 0$ such that $|h| < \delta$ implies*

$$\sup_{f \in D} \|\tau_h f - f\|_{L^p(\mathbb{R}^d)} < \varepsilon,$$

where $\tau_f f = f(\cdot + h)$. Then $D\lceil_U$ is pre-compact in $L^p(\mathbb{R}^d)$ for any measurable set $U \subset \mathbb{R}^d$ with finite measure. Here $D\lceil_U = \{f\lceil_U | f \in D\}$.

Proof The proof is taken from [11, Theorem 4.26]. It suffices to show that $D\lceil_U$ is totally bounded,[3] since $L^p(U)$ is complete.

Let $\rho_n \in C_0^\infty(\mathbb{R}^d)$ be such that

$$\rho_n(x) \ge 0, \quad \mathrm{supp}\rho_n \subset \overline{B(0, 1/n)}, \quad \int_{\mathbb{R}^d} \rho_n(x)dx = 1.$$

Here $\overline{B(0, 1/n)}$ denotes the closed ball centered at the origin with radius $1/n$. Then we have

$$\begin{aligned} |(\rho_n * f)(x) - f(x)| &\le \int_{\mathbb{R}^d} |f(x-y) - f(x)|\rho_n(y)dy \\ &\le \left(\int_{\mathbb{R}^d} |f(x-y) - f(x)|^p \rho_n(y)dy \right)^{1/p} \end{aligned}$$

by Hölder inequality. By assumption we obtain

$$\begin{aligned} &\int_{\mathbb{R}^d} |(\rho_n * f)(x) - f(x)|^p dy \le \int_{\mathbb{R}^d}\int_{\mathbb{R}^d} |f(x-y) - f(x)|^p \rho_n(y)dxdy \\ &= \int_{B(0,1/n)} \rho_n(y)dy \int_{\mathbb{R}^d} |f(x-y) - f(x)|^p dx \le \varepsilon^p \end{aligned}$$

for any $f \in D$ provided $1/n < \delta$. Then it follows that

$$\sup_{f \in D} \|(\rho_n * f) - f\|_{L^p(\mathbb{R}^d)} \le \varepsilon$$

---

[2] D is uniformly equicontinuous if and only if for $\forall \varepsilon > 0$ there exists $\delta > 0$ such that $d(x, y) < \delta$ implies $\sup_{f \in D} |f(x) - f(y)| < \varepsilon$.

[3] X is totally bounded if and only if for given any $\varepsilon > 0$ there is a finite covering of X by balls of radius ε.

for any $n > 1/\delta$. We also have

$$\sup_{f \in D} \|\rho_n * f\|_{L^\infty(\mathbb{R}^d)} \le C_n \sup_{f \in D} \|f\|_{L^p(\mathbb{R}^d)} < \infty \tag{2.2}$$

by Hölder inequality. Let $p^* = dp/(d-p)$. Since $\|\nabla(\rho_n * f)\|_{L^\infty(\mathbb{R}^d)} \le C'_n \|\nabla \rho_n\|_{p^*} \|f\|_{L^p(\mathbb{R}^d)}$, we have

$$|(\rho_n * f)(x) - (\rho_n * f)(y)| \le C''_n \|f\|_{L^p(\mathbb{R}^d)} |x - y|.$$

for any $f \in D$ and any $x, y \in \mathbb{R}^d$. Thus

$$\sup_{f \in D} |(\rho_n * f)(x) - (\rho_n * f)(y)| \le C'''_n |x - y| \tag{2.3}$$

for any $x, y \in \mathbb{R}^d$. Equations (2.2) and (2.3) imply that set $\{\rho_n * f \mid f \in D\}$ is bounded and equicontinuous. Note that U is not necessarily bounded. We will approximate U by some bounded set V. We have

$$\|f\|_{L^p(U \setminus V)} \le \|f - (\rho_n * f)\|_{L^p(\mathbb{R}^d)} + \|\rho_n * f\|_{L^p(U \setminus V)}$$

and

$$\|\rho_n * f\|_{L^p(U \setminus V)} \le \|\rho_n * f\|_{L^\infty(\mathbb{R}^d)} |U \setminus V|.$$

Here $|U \setminus V|$ is the measure of $U \setminus V$. Choose a bounded measurable V such that $|U \setminus V| < \varepsilon/2$ and take sufficiently large n such that $\|f - (\rho_n * f)\|_{L^p(\mathbb{R}^d)} < \varepsilon/2$. We have

$$\sup_{f \in D} \|f\|_{L^p(U \setminus V)} < \varepsilon. \tag{2.4}$$

Fix V and $n > 1/\delta$ such that (2.4) holds. Hence

$$\mathscr{D} = (\rho_n * D)\lceil_{\bar{V}} = \{(\rho_n * f)\lceil_{\bar{V}} \mid f \in D\}$$

satisfies assumptions of the Ascoli-Arzelà theorem. Then $\mathscr{D}$ has compact closure in $C(\bar{V})$. This implies that $\mathscr{D}$ also has compact closure in $L^p(V)$. Hence we can cover $\mathscr{D}$ by a finite number of balls of radius ε in $L^p(V)$:

$$\mathscr{D} \subset \cup_{j=1}^N B(g_j, \varepsilon), \quad g_j \in L^p(V) \tag{2.5}$$

with some $N < \infty$. Define

$$\bar{g}_j = \begin{cases} g_j & \text{on } V, \\ 0 & \text{on } U \setminus V. \end{cases}$$

Let $f \in D$. There exists g_j such that $\rho_n * f \in B(g_j, \varepsilon)$ by (2.5). Since

$$\|f-\bar{g}_j\|_{L^p(U)}^p=\int_{U\setminus V}|f|^p dx+\int_V |f-g_j|^p dx,$$

we have

$$\begin{aligned}\|f-\bar{g}_j\|_{L^p(U)} &\le \varepsilon+\|f-g_j\|_{L^p(V)}\\ &\le \varepsilon+\|f-(\rho_n * f)\|_{L^p(\mathbb{R}^d)}+\|(\rho_n * f)-g_j\|_{L^p(V)}<3\varepsilon.\end{aligned}$$

We conclude that $f\in B(\bar{g}_j,3\varepsilon)$ and

$$D\lceil_U \subset \cup_{j=1}^N B(\bar{g}_j,3\varepsilon).$$

Then $D\lceil_U$ has compact closure in $L^p(U)$. ■

Corollary 2.14 *Let $1\le p<\infty$. Let $D\subset L^p(\mathbb{R}^d)$ be bounded. Suppose the same assumptions as in Theorem 2.13. In addition we suppose that for arbitrary $\varepsilon>0$, there exists $U\subset\mathbb{R}^d$ such that U is bounded and $\|f\|_{L^p(\mathbb{R}^d\setminus U)}<\varepsilon$ for all $f\in D$. Then D has compact closure in $L^p(\mathbb{R}^d)$.*

Proof By Theorem 2.13, $D\lceil_U$ has compact closure in $L^p(U)$. Then

$$D\lceil_U \subset \cup_{j=1}^N B(g_j,\varepsilon),\quad g_j\in L^p(U).$$

Set $\bar{g}_j=\begin{cases} g_j & \text{on } U,\\ 0 & \text{on } \mathbb{R}^d\setminus U.\end{cases}$ Let $f\in D$. There exists g_j such that $\|f\lceil_U-g_j\|_{L^p(\mathbb{R}^d)}<\varepsilon$. We have $\|f-g_j\|_{L^p(\mathbb{R}^d)}\le\|f\lceil_U-g_j\|_{L^p(\mathbb{R}^d)}+\|f\lceil_{U^c}\|_{L^p(\mathbb{R}^d)}<2\varepsilon$. This implies that subset D is covered by balls $B(\bar{g}_j,2\varepsilon)$, $1=1,\ldots,N$, in $L^p(\mathbb{R}^d)$. ■

2.2.2 *Sobolev Inequalities*

Our next goal is to discover embeddings of various Sobolev spaces into others. The crucial tool is Sobolev inequalities.

Let D denote the gradient operator: $Df=(\partial_{x_1}f,\ldots,\partial_{x_d}f)$ and $D^\alpha=\partial_{x_1}^{\alpha_1}\cdots\partial_{x_d}^{\alpha_d}$ for $\alpha=(\alpha_1,\ldots,\alpha_d)$. Suppose that $f,g\in L^1_{\rm loc}(U)$, $U\subset\mathbb{R}^d$. We say that g is the α_{th} weak-partial derivative of f, written as $D^\alpha f=g$, provided that

$$\int_U f(x)D^\alpha\phi(x)dx=(-1)^{|\alpha|}\int_U g(x)\phi(x)dx$$

holds for any $\phi\in C_0^\infty(U)$. Here $|\alpha|=\sum_{j=1}^d\alpha_j$. The Sobolev space $W^{k,p}(U)$ for $1\le p\le\infty$ and nonnegative integer k is defined by

$$W^{k,p}(U)=\left\{f\in L^1_{\rm loc}(U)|D^\alpha f\in L^p(U)\text{ for all }\alpha\text{ with }0\le|\alpha|\le k\right\}.$$

Introduce the norm

$$\|f\|_{W^{k,p}(U)} = \left(\sum_{0 \le |\alpha| \le k} \int_U |D^\alpha f|^p dx \right)^{1/p}, \quad 1 \le p < \infty$$

on $W^{k,p}(U)$. By this norm $W^{k,p}(U)$ turns to be a Banach space.

The purpose of this section is to see that embedding

$$\iota : W^{1,p}(U) \to L^q(U), \quad 1 \le q < \frac{dp}{d-p}$$

is a compact operator. Hence bounded sequence $(f_n)_{n=1}^{\infty} \subset W^{1,p}(U)$ contains a strongly convergent subsequence $(f_{n_k})_{k=1}^{\infty}$ in $L^q(U)$.

We can see that $C_0^\infty(U) \subset L^p(U) \subset W^{k,p}(U)$. Note that

$$\overline{C_0^\infty(\mathbb{R}^d)}^{\|\cdot\|_{W^{k,p}(\mathbb{R}^d)}} = W^{k,p}(\mathbb{R}^d)$$

but

$$\overline{C_0^\infty(U)}^{\|\cdot\|_{W^{k,p}(U)}} \neq W^{k,p}(U)$$

in general for $U \subsetneq \mathbb{R}^d$. To approximate $f \in W^{k,p}(U)$ by functions in $C_0^\infty(U)$ we need an extension of f to $\tilde{f} \in W^{k,p}(\mathbb{R}^d)$, since $\overline{C_0^\infty(\mathbb{R}^d)}^{\|\cdot\|_{W^{k,p}(\mathbb{R}^d)}} = W^{k,p}(\mathbb{R}^d)$ implies that $\tilde{f}$ can be approximated by some functions in $C_0^\infty(\mathbb{R}^d)$. Therefore it is very important to extend a function in $W^{1,p}(U)$ to some function in $W^{1,p}(\mathbb{R}^d)$, which is however not straightforward. If we extend $f \in W^{1,p}(U)$ to

$$\tilde{f}(x) = \begin{cases} f(x), & x \in U, \\ 0, & x \notin U, \end{cases}$$

$\tilde{f} \notin W^{1,p}(\mathbb{R}^d)$ in general, since we may create such a bad discontinuity along the boundary ∂U that $\tilde{f}$ no longer has weak derivatives.

Let $U, V \subset \mathbb{R}^d$. We write $U \subset\subset V$ if and only if $U \subset \bar{U} \subset V$ holds and $\bar{U}$ is compact.

Proposition 2.15 (Extension theorem) *Suppose that $U \subset \mathbb{R}^d$ is bounded and ∂U is C^1. Let V be a bounded open set such that $U \subset\subset V$. Then there exists a bounded linear operator $E : W^{1,p}(U) \to W^{1,p}(\mathbb{R}^d)$ such that (1) $Ef = f$ a.e. on U, (2)* supp *$Ef \subset V$ and (3) $\|Ef\|_{W^{1,p}(\mathbb{R}^d)} \le C\|f\|_{W^{1,p}(U)}$, where $C = C(p, U, V)$ is a constant.*

Proof See e.g., [14, p.254, Theorem 1]. ■

Simply we set $\|f\|_{L^p(\mathbb{R}^d)} = \|f\|_p$. We are interested in establishing the inequality of the form: $\|f\|_q \le C\|Df\|_p$, but q and p are restricted by a scaling argument.

Introducing the scaling $f_\lambda(x) = f(\lambda x)$, we insert f_λ into $\|f\|_q \le C\|Df\|_p$ and get

$$\|f\|_q \le C\lambda^{1-\frac{d}{p}+\frac{d}{q}}\|Df\|_p.$$

Then $1 - \frac{d}{p} + \frac{d}{q} = 0$ should be satisfied to get the inequality. Then we set

$$p^* = \frac{dp}{d-p}$$

and it satisfies that $1 - \frac{d}{p} + \frac{d}{p^*} = 0$.

Now we introduce two inequalities according to $p > d$ and $p < d$. We say $f \in C_u^k(U)$ if and only if $f \in C^k(U)$ and $D^\alpha f$ is uniformly continuous on any bounded subsets of U for each α with $|\alpha| \le k$, and the γ-Hölder semi-norm on $C_u^k(U)$ is defined by

$$\|f\|_{C_u^{0,\gamma}(U)} = \sup_{x,y\in U, x\neq y} \frac{|f(x)-f(y)|}{|x-y|^\gamma}.$$

We set

$$C_u^{k,\gamma}(U) = \left\{ f \in C_u^k(U) \,\middle|\, \|f\|_{C_u^{k,\gamma}(U)} = \sum_{|\alpha|\le k} \|D^\alpha f\|_{L^\infty(U)} + \sum_{|\alpha|=k} \|D^\alpha f\|_{C_u^{0,\gamma}(U)} < \infty \right\}$$

and $C_u^{k,\gamma}(U)$ is called a Hölder space and is a Banach space. We have propositions below.

Proposition 2.16 (Cagliardo–Nirenberg–Sobolev inequality) *Let $1 \le p < d$. Then there exists a constant $C = C(p,d)$ such that $\|f\|_{p^*} \le C\|Df\|_p$ for any $f \in C_0^1(\mathbb{R}^d)$.*

Proposition 2.17 (Morrey's inequality) *Let $d < p \le \infty$. Then there exists a constant $C = C(p,d)$ such that $\|f\|_{C_u^{0,1-\frac{d}{p}}(\mathbb{R}^d)} \le C\|f\|_{W^{1,p}(\mathbb{R}^d)}$ for any $f \in C_0^1(\mathbb{R}^d)$.*

See e.g., [14] for proofs of Propositions 2.16 and 2.17. From Propositions 2.16 and 2.17 we can estimate functions in $W^{1,p}(U)$.

Proposition 2.18 (Sobolev inequality) *Let U be bounded and open in $\mathbb{R}^d$, and ∂U is C^1. Let $f \in W^{1,p}(U)$.*

(1) Let $1 \le p < d$. Then $f \in L^{p^*}(U)$ and $\|f\|_{L^{p^*}(U)} \le C\|f\|_{W^{1,p}(U)}$.

(2) Let $d < p \le \infty$. Then $f \in C_u^{0,1-\frac{d}{p}}(U)$ and $\|f\|_{C_u^{0,1-\frac{d}{p}}(U)} \le C\|f\|_{W^{1,p}(U)}$.

Proof Consider (1). There exists an extension $\tilde{f} = Ef \in W^{1,p}(\mathbb{R}^d)$ of f such that $\tilde{f}$ has a compact support and $\|\tilde{f}\|_{W^{1,p}(\mathbb{R}^d)} \le C\|f\|_{W^{1,p}(\mathbb{R}^d)}$. There exist functions $\{f_n\}_{n=1}^\infty \subset C_0^\infty(\mathbb{R}^d)$ such that $f_n \to \tilde{f}$ in $W^{1,p}(\mathbb{R}^d)$. By Cagliardo–Nirenberg–Sobolev inequality we can see that $\{f_n\}_{n=1}^\infty$ is a Cauchy sequence in $L^{p^*}(\mathbb{R}^d)$ and $\|f_n\|_{p^*} \le C\|f_n\|_{W^{1,p}(\mathbb{R}^d)}$. Then (1) follows from a limiting argument.

Next consider (2). Similarly to (1) let $Ef = \tilde{f} \in W^{1,p}(\mathbb{R}^d)$ be an extension, and $\{f_n\}_{n=1}^{\infty} \subset C_0^{\infty}(\mathbb{R}^d)$ satisfies that $f_n \to \tilde{f}$ in $W^{1,p}(\mathbb{R}^d)$. By Morrey's inequality, we can also see that $(f_n)_{n=1}^{\infty}$ is Cauchy in $C_u^{0,1-\frac{d}{p}}(\mathbb{R}^d)$ and there exists $f^* \in C_u^{0,1-\frac{d}{p}}(\mathbb{R}^d)$ such that $f_n \to f^*$ in $C_u^{0,1-\frac{d}{p}}(\mathbb{R}^d)$. Thus it follows that $f^* = f$ a.e. on U. From inequality $\|f_n\|_{C_u^{0,1-\frac{d}{p}}(\mathbb{R}^d)} \le C\|f_n\|_{W^{1,p}(\mathbb{R}^d)}$ and a limiting argument (2) follows. ■

We can extend these results as follows.

Proposition 2.19 (Sobolev inequality) *Let U be bounded and open in $\mathbb{R}^d$, and ∂U is C^1. Assume that $f \in W^{k,p}(U)$. Then (1) and (2) follow.*

(1) Let $k < d/p$. Then $f \in L^q(U)$ with $\frac{1}{q} = \frac{1}{p} - \frac{k}{d}$ and

$$\|f\|_{L^q(U)} \le C\|f\|_{W^{k,p}(U)}.$$

(2) Let $k > d/p$. Then $f \in C^{k-[d/p]-1,\gamma}(U)$ and

$$\|f\|_{C_u^{k-[d/p]-1,\gamma}(U)} \le C\|f\|_{W^{1,p}(U)},$$

where

$$\gamma = \begin{cases} [\frac{d}{p}] + 1 - \frac{d}{p}, & \frac{d}{p} \notin \mathbb{Z}, \\ \forall a \in \mathbb{R} \text{ such that } 0 < a < 1, & \frac{d}{p} \in \mathbb{Z}. \end{cases}$$

Proof See e.g., [14]. ■

2.2.3 Compact Embeddings

Let X and Y be Banach spaces such that $X \subset Y$. We say that X is compactly embedded in Y provided that (1) $\|f\|_Y \le C\|f\|_X$ for $f \in X$ with some constant C, (2) for arbitrary bounded sequence $\{f_n\}_{n=1}^{\infty} \subset X$ is pre-compact in Y. When X is compactly embedded in Y, we write $X \subset\subset Y$. In (1) of Proposition 2.18 we see that $W^{1,p}(U)$ can be *continuously* embedded in $L^{p^*}(U)$. In this section we show that $W^{1,p}(U)$ can be *compactly* embedded in $L^q(U)$ for $1 \le q < p^*$.

Theorem 2.20 (Rellich–Kondrachov compactness theorem) *Let U be bounded and open in $\mathbb{R}^d$, and ∂U be C^1. Let $1 \le p < d$. Then $W^{1,p}(U) \subset\subset L^q(U)$ for each $1 \le q < p^*$.*

Proof We take the proof below from [14, Theorem 1, p.272]
(Step 1) Since $W^{1,p}(U) \subset L^q(U)$ and $\|f\|_{L^q(U)} \le C\|f\|_{W^{1,p}(U)}$, it remains to show that if $\{f_n\}_{n=1}^{\infty} \subset W^{1,p}(U)$ is a bounded sequence, there exists a convergent subsequence $\{f_{n_k}\}_{k=1}^{\infty}$ in $L^q(U)$. By the extension theorem (Theorem 2.15) without loss

of generality we can assume that $U = \mathbb{R}^d$, and all f_n have compact support in some bounded open set $V \subset \mathbb{R}^d$. Furthermore $K = \sup_{n\in\mathbb{N}} \|f_n\|_{W^{1,p}(V)} < \infty$.
(Step 2) Let $f_n^\varepsilon = f_n * \rho_\varepsilon \in C_0^\infty(V)$. Here $0 \le \rho \in C_0^\infty(\mathbb{R}^d)$ such that $\int_{\mathbb{R}^d} \rho(x)dx = 1$ and $\rho_\varepsilon(x) = \rho(x/\varepsilon)\varepsilon^{d/2}$. We shall show that $f_n^\varepsilon \to f_n$ in $L^q(V)$ uniformly in n in this step. We have $\int_V |f_n^\varepsilon(x) - f_n(x)|dx \le \varepsilon \int_V |Df_n(x)|dx$. Hence

$$\|f_n^\varepsilon - f_n\|_{L^1(V)} \le \varepsilon\|Df_n\|_{L^1(V)} \le \varepsilon C\|Df_n\|_{L^p(V)} \le \varepsilon C\|f_n\|_{W^{1,p}(V)} \le \varepsilon K.$$

This implies that $f_n^\varepsilon \to f_n$ in $L^1(V)$ uniformly in n. By the interpolation we have

$$\|f_n^\varepsilon - f_n\|_{L^q(V)} \le \|f_n^\varepsilon - f_n\|_{L^1(V)}^\theta \|f_n^\varepsilon - f_n\|_{L^r(V)}^{1-\theta},$$

where $\frac{1}{q} = \theta + \frac{1-\theta}{r}$ and $0 < \theta < 1$. By Cagliardo–Nirenberg–Sobolev inequality we have

$$\|f_n^\varepsilon - f_n\|_{L^q(V)} \le C\|f_n^\varepsilon - f_n\|_{L^1(V)}^\theta.$$

Thus $f_n^\varepsilon \to f_n$ in $L^q(V)$ uniformly in n.
(Step 3) It is straightforward to see that $|f_n^\varepsilon(x)| \le \|\rho_\varepsilon\|_\infty \|f_n\|_{L^1(V)} \le C/\varepsilon^d$ and similarly $|Df_n^\varepsilon(x)| \le \|D\rho_\varepsilon\|_\infty \|f_n\|_{L^1(V)} \le C/\varepsilon^{d+1}$. Then $\{f_n^\varepsilon\}_{n=1}^\infty$ is uniformly bounded and equicontinuous for each $\varepsilon > 0$.[4]
(Step 4) Fix $\delta > 0$. Take $\varepsilon > 0$ such that $\|f_n^\varepsilon - f_n\|_{L^q(V)} < \delta$ for any n. Since f_n^ε, $n = 1, 2, \ldots$, have support in bounded set V, by Ascoli–Arzela theorem there exists a subsequence $\{f_{n_k}^\varepsilon\}_{k=1}^\infty$ converging uniformly on V. Note that subsequence $\{f_{n_k}^\varepsilon\}_{k=1}^\infty$ depends on the choice of ε and δ. Since

$$\|f_{n_k} - f_{n_j}\|_{L^q(V)} \le \|f_{n_k} - f_{n_k}^\varepsilon\|_{L^q(V)} + \|f_{n_k}^\varepsilon - f_{n_j}^\varepsilon\|_{L^q(V)} + \|f_{n_j}^\varepsilon - f_{n_j}\|_{L^q(V)},$$

we have $\limsup_{j,k\to\infty} \|f_{n_k} - f_{n_j}\|_{L^q(V)} \le \delta + \limsup_{j,k\to\infty} \|f_{n_k}^\varepsilon - f_{n_j}^\varepsilon\|_{L^q(V)} \le \delta$. We can conclude that for a given $\delta > 0$ there exists a subsequence $\{f_{n_k}^\varepsilon\}_{k=1}^\infty$ such that

$$\limsup_{j,k\to\infty} \|f_{n_k} - f_{n_j}\|_{L^q(V)} \le \delta. \tag{2.6}$$

We employ (2.6) with $\delta = 1, 1/2, 1/3, \ldots$. By a standard diagonal argument we can extract a subsequence $\{m_l\}$ such that $\limsup_{l,k\to\infty} \|f_{m_k} - f_{m_l}\|_{L^q(V)} = 0$. Then the proposition is proven. ■

Remark 2.21 Let $U \subset \mathbb{R}^d$ be bounded and open with ∂U being C^1. In the case of $p = d$, we have $W^{1,p}(U) \subset\subset L^q(U)$ for any $q \in [p, \infty)$. This follows from the fact $W^{1,p}(U) \subset W^{1,p'}(U)$ for $p' < p$ and Rellich–Kondrachov compactness theorem. In the case of $p > d$, $W^{1,p}(U) \subset\subset C(\bar{U})$. This follows from Proposition 2.18 and Ascoli–Arzela theorem.

---

[4]Note that $F \subset C(X)$ is equicontinuous on metric space (X, d) if and only if for every $\varepsilon > 0$ there exists $\delta > 0$ such that for $x, y \in X, d(x, y) < \delta$ implies that $\|f(x) - f(y)\| < \varepsilon$ for arbitrary $f \in F$.

2.3 Boson Fock Space

2.3.1 Abstract Boson Fock Spaces

In this section we prepare fundamental tools used in quantum field theory. Quantum field theory treats particles as excited states of their underlying fields, which are more fundamental than classical particles, and particles are created and annihilated, and the number of particles changes. To describe creation and annihilation of particles we need a Fock space.

Let $\mathscr{W}$ be a separable Hilbert space over the complex field $\mathbb{C}$. The choice of Hilbert space $\mathscr{W}$ depends on the species of particle we consider. When we are describing a system of n particles of the same species, we could take the Hilbert space

$$\otimes^n \mathscr{W} = \underbrace{\mathscr{W} \otimes \cdots \otimes \mathscr{W}}_{n},$$

which is however too big. It is known that boson particles of the same species are indistinguishable. Hence there is no way to attach labels to any two particles so as to tell which one is which. So we consider only symmetric state $f_1 \otimes \cdots \otimes f_n$ such that $f_{\pi(1)} \otimes \cdots \otimes f_{\pi(n)} = f_1 \otimes \cdots \otimes f_n$ for any $\pi \in \wp_n$. Here $\wp_n$ denotes the permutation group of n letters. Taking into account the above argument we consider the operation $\otimes_{\mathrm{s}}^n$ of n-fold symmetric tensor product of $\mathscr{W}$ defined through the symmetrization operator

$$S_n(f_1 \otimes \cdots \otimes f_n) = \frac{1}{n!} \sum_{\pi \in \wp_n} f_{\pi(1)} \otimes \cdots \otimes f_{\pi(n)}, \quad n \geq 1.$$

We extend S_n to the linear operator on $\otimes^n \mathscr{W}$. Define the n-fold symmetric tensor product of $\mathscr{W}$ by

$$\otimes_{\mathrm{s}}^n \mathscr{W} = S_n(\otimes^n \mathscr{W}),$$

where we set $\otimes_{\mathrm{s}}^0 \mathscr{W} = \mathbb{C}$ as usual. In order to form a state space which can accommodate any number of identical boson particles, we put all $\otimes_{\mathrm{s}}^n \mathscr{W}$ together to form the complete symmetric tensor algebra over $\mathscr{W}$.

Definition 2.22 (*Boson Fock space*) The boson Fock space over $\mathscr{W}$ is the Hilbert space defined by the infinite direct sum of $\otimes_{\mathrm{s}}^n \mathscr{W}$

$$\mathscr{F}(\mathscr{W}) = \oplus_{n=0}^{\infty} \left[\otimes_{\mathrm{s}}^n \mathscr{W}\right],$$

endowed with the scalar product

$$(\Psi, \Phi)_{\mathscr{F}(\mathscr{W})} = \sum_{n=0}^{\infty} (\Psi^{(n)}, \Phi^{(n)})_{\otimes_{\mathrm{s}}^n \mathscr{W}}.$$

We simply denote $\mathscr{F}(\mathscr{W})$ by $\mathscr{F}$, and $\otimes_s^n \mathscr{W}$ by $\mathscr{F}^{(n)}$. $\mathscr{F}^{(n)}$ is called the n-particle subspace of $\mathscr{F}$.

Thus $\mathscr{F}$ is a state space which describes arbitrary numbers of identical bosons including none at all. The boson Fock space $\mathscr{F}$ can be identified with the space of ℓ_2-sequences $(\Psi^{(n)})_{n=0}^{\infty}$ such that $\Psi^{(n)} \in \otimes_s^n \mathscr{W}$ for $n \geq 0$. The element

$$\Omega = (1, 0, 0, \dots) \in \mathscr{F}$$

is called the Fock vacuum. It is convenient to consider the finite particle subspace of states in which the total number of particles is bounded above:

$$\mathscr{F}_{\text{fin}} = \left\{ (\Psi^{(n)})_{n=0}^{\infty} \in \mathscr{F} \,\middle|\, \Psi^{(m)} = 0 \text{ for all } m \geq M \text{ with some } M \right\}.$$

We now introduce some operators that raise or lower the number of particles. The creation operator $a^\dagger(f)$, $f \in \mathscr{W}$, is defined by

$$(a^\dagger(f)\Psi)^{(n)} = \begin{cases} \sqrt{n} S_n(f \otimes \Psi^{(n-1)}), & n \geq 1, \\ 0, & n = 0 \end{cases}$$

with domain

$$D(a^\dagger(f)) = \left\{ (\Psi^{(n)})_{n=0}^{\infty} \in \mathscr{F} \,\middle|\, \sum_{n=1}^{\infty} n \| S_n(f \otimes \Psi^{(n-1)}) \|_{\mathscr{F}^{(n)}}^2 < \infty \right\},$$

and the annihilation operator $a(f)$ by

$$a(f) = (a^\dagger(\bar{f}))^*, \quad f \in \mathscr{W}.$$

In what follows $a^\sharp$ describes a or $a^\dagger$. The action of $a^\dagger(f)$ creates a particle in the state f, whereas $a(f)$ destroys it. Furthermore we can see that $a(f)$ annihilates any multi-particle state in which no particle has any probability of being in the state of f, i.e., $a(f) \prod_{j=1}^{n} a^\dagger(g_j)\Omega = 0$ if $(f, g_j) = 0$ for $j = 1, ..., n$. Since one is the adjoint operator of the other, the relation $(\Phi, a(f)\Psi)_{\mathscr{F}} = (a^\dagger(\bar{f})\Phi, \Psi)_{\mathscr{F}}$ holds. Furthermore, since both operators are closable, we denote their closed extensions by the same symbols in what follows.

Remark 2.23 In some literatures on quantum field theory annihilation operator $a(f)$ is defined by $a(f) = (a^\dagger(f))^*$. This means that $a(f)$ is anti-linear with respect to f. On the contrary in our notation annihilation operator $a(f)$ is linear in f.

Let $D \subset \mathscr{W}$ be a dense subspace of $\mathscr{W}$. It is known that

$$\mathscr{F}^{(n)} = \overline{\text{L.H.}\{a^\dagger(f_1) \cdots a^\dagger(f_n)\Omega \,|\, f_j \in D, j = 1, .., n\}}, \quad n \geq 1, \tag{2.7}$$

where $\overline{\{\cdots\}}$ denotes the closure in $\mathscr{F}$. Equation (2.7) means that the n-particle subspace $\mathscr{F}^{(n)}$ is created by n-creation operators and the Fock vacuum. The operators a and $a^\dagger$ leave $\mathscr{F}_{\text{fin}}$ invariant. Thus we can do algebraic computations of a and $a^\dagger$ on $\mathscr{F}_{\text{fin}}$ without worrying about operator domains. The most important consequence of introducing normalization factor $\sqrt{n}$ into the definition of $a^\dagger(f)$ makes commutation relations of $a^\sharp(f)$ simple:

$$[a(f), a^\dagger(g)] = (\bar{f}, g)_{\mathscr{W}} \mathbb{1}, \quad [a(f), a(g)] = 0, \quad [a^\dagger(f), a^\dagger(g)] = 0.$$

This commutation relations are called canonical commutation relations.

Next we discuss canonical identification of $\mathscr{F}(\mathscr{W}_1 \oplus \mathscr{W}_2)$ and $\mathscr{F}(\mathscr{W}_1) \otimes \mathscr{F}(\mathscr{W}_2)$. An example is $\mathscr{W} = L^2(\mathbb{R}^d) = L^2(|k| < \Lambda) \oplus L^2(|k| \geq \Lambda) = \mathscr{W}_1 \oplus \mathscr{W}_2$. Let Ω, Ω_1 and Ω_2 be Fock vacuum of $\mathscr{F}(\mathscr{W})$, $\mathscr{F}(\mathscr{W}_1)$ and $\mathscr{F}(\mathscr{W}_2)$, respectively. Let us consider two Hilbert spaces $\mathscr{W}_1$ and $\mathscr{W}_2$. $\mathscr{W} = \mathscr{W}_1 \oplus \mathscr{W}_1$ is also a Hilbert space with scalar product $(F, G) = (f_1, g_1)_{\mathscr{W}_1} + (f_2, g_2)_{\mathscr{W}_2}$ for $F = f_1 \oplus f_2$ and $G = g_1 \oplus g_2$, and we can define the boson Fock space over $\mathscr{W}$, $\mathscr{F}(\mathscr{W})$.

Let us consider the map $\mathscr{U}_{\text{c}} : \mathscr{F}(\mathscr{W}) \to \mathscr{F}(\mathscr{W}_1) \otimes \mathscr{F}(\mathscr{W}_2)$ by

$$\mathscr{U}_{\text{c}} \prod_{j=1}^{n} a^\dagger(f_j \oplus 0) \prod_{i=1}^{m} a^\dagger(0 \oplus g_i)\Omega = \prod_{j=1}^{n} a^\dagger(f_j)\Omega_1 \otimes \prod_{i=1}^{m} a^\dagger(g_i)\Omega_2.$$

$\mathscr{U}_{\text{c}}$ maps $\mathscr{F}(\mathscr{W})^{(n+m)}$ to $\mathscr{F}(\mathscr{W}_1)^{(n)} \otimes \mathscr{F}(\mathscr{W}_2)^{(m)}$. We extend $\mathscr{U}_{\text{c}}$ to the map from $\mathscr{F}_{\text{fin}}(\mathscr{W})$ to $\mathscr{F}_{\text{fin}}(\mathscr{W}_1) \otimes \mathscr{F}_{\text{fin}}(\mathscr{W}_2)$ by the linearity. We can also check that $\mathscr{U}_{\text{c}}$ is isometry and surjective. Thus we can extend $\mathscr{U}_{\text{c}}$ to the unitary operator from $\mathscr{F}(\mathscr{W})$ to $\mathscr{F}(\mathscr{W}_1) \otimes \mathscr{F}(\mathscr{W}_2)$ and we establish the identification

$$\mathscr{F}(\mathscr{W}_1 \oplus \mathscr{W}_2) \cong \mathscr{F}(\mathscr{W}_1) \otimes \mathscr{F}(\mathscr{W}_2).$$

Recursively we can also see the unitary equivalence

$$\mathscr{F}(\oplus_{j=1}^{n} \mathscr{W}_j) \cong \mathscr{F}(\mathscr{W}_1) \otimes \cdots \otimes \mathscr{F}(\mathscr{W}_n).$$

2.3.2 *Field Operators and Second Quantizations*

The creation operator and the annihilation operator are not symmetric and do not commute with each others. We can, however, construct a family of symmetric and commutative operators by linearly combining the operators and this leads to the so-called field operators. The field operator $\Phi(f)$ is defined by

$$\Phi(f) = \frac{1}{\sqrt{2}}(a^\dagger(f) + a(\bar{f})), \quad f \in \mathscr{W},$$

and its conjugate momentum by

$$\Pi(f) = \frac{i}{\sqrt{2}}(a^\dagger(f) - a(\bar{f})), \quad f \in \mathscr{W}.$$

Both $\Phi(f)$ and $\Pi(g)$ are symmetric, however, not linear in f and g over the complex field $\mathbb{C}$. Note that, in contrast, they are linear operators over the real field $\mathbb{R}$. Using canonical commutation relations of creation operators and annihilation operators we have

$$[\Phi(f), \Pi(g)] = i\mathrm{Re}(f, g), \quad [\Phi(f), \Phi(g)] = i\mathrm{Im}(f, g), \quad [\Pi(f), \Pi(g)] = i\mathrm{Im}(f, g).$$

In particular, for real-valued f and g,

$$[\Phi(f), \Pi(g)] = i(f, g), \quad [\Phi(f), \Phi(g)] = [\Pi(f), \Pi(g)] = 0. \tag{2.8}$$

Proposition 2.24 (Nelson's analytic vector theorem [48]) *Let K be a symmetric operator on a Hilbert space. Assume that there exists a dense subspace $\mathscr{D} \subset D(K)$ such that $\lim_{m\to\infty} \sum_{n=0}^{m} \|K^n f\| t^n/n! < \infty$, for $f \in \mathscr{D}$ and some $t > 0$. Then K is essentially selfadjoint on $\mathscr{D}$, and*

$$e^{-tK}\Phi = s - \lim_{m\to\infty} \sum_{n=0}^{m} \frac{t^n K^n}{n!} f$$

follows for $f \in \mathscr{D}$.

We see that $\mathscr{F}_{\mathrm{fin}}$ is the set of analytic vectors of $\Phi(f)$ and $\Pi(g)$, i.e.,

$$\lim_{m\to\infty} \sum_{n=0}^{m} \frac{\|\Phi(f)^n \Psi\| t^n}{n!} < \infty, \quad \lim_{m\to\infty} \sum_{n=0}^{m} \frac{\|\Pi(g)^n \Psi\| t^n}{n!} < \infty$$

for $\Psi \in \mathscr{F}_{\mathrm{fin}}$ and $t \geq 0$. By Nelson's analytic vector theorem both $\Phi(f)$ and $\Pi(g)$ are essentially selfadjoint on $\mathscr{F}_{\mathrm{fin}}$. We keep denoting the closures of $\Phi(f)\lceil_{\mathscr{F}_{\mathrm{fin}}}$ and $\Pi(g)\lceil_{\mathscr{F}_{\mathrm{fin}}}$ by the same symbols. We define unitary operators $e^{it\Phi(f)}$ and $e^{is\Pi(g)}$. They satisfy Weyl relation:

$$e^{it\Phi(f)} e^{is\Pi(g)} = e^{-ist\mathrm{Re}(f,g)} e^{is\Pi(g)} e^{it\Phi(f)}, \quad s, t \in \mathbb{R}.$$

We define the second quantization $\Gamma(T)$ of contraction operator T and the differential second quantization $\mathrm{d}\Gamma(h)$ of selfadjoint operator h. Given a contraction operator T on $\mathscr{W}$, the second quantization of T is the contraction operator $\Gamma(T)$ on $\mathscr{F}$ defined by

$$\Gamma(T) = \oplus_{n=0}^{\infty}[\otimes^n T].$$

Here it is understood that $\otimes^0 T = \mathbb{1}$. $\Gamma(T)$ acts as

$$\Gamma(T)\prod_{j=1}^{n} a^\dagger(f_j)\Omega = \prod_{j=1}^{n} a^\dagger(Tf_j)\Omega.$$

The map Γ satisfies semigroup properties:

$$\Gamma(S)\Gamma(T) = \Gamma(ST), \quad \Gamma(S)^* = \Gamma(S^*), \quad \Gamma(\mathbb{1}_{\mathscr{W}}) = \mathbb{1}_{\mathscr{F}}. \tag{2.9}$$

For a selfadjoint operator h on $\mathscr{W}$ the structure relations (2.9) imply in particular that $\{\Gamma(e^{ith})\}_{t\in\mathbb{R}}$ is a strongly continuous one-parameter unitary group on $\mathscr{F}$. By Stone's theorem[5] there exists a unique selfadjoint operator $\mathrm{d}\Gamma(h)$ on $\mathscr{F}$ such that

$$\Gamma(e^{ith}) = e^{it\mathrm{d}\Gamma(h)}, \quad t \in \mathbb{R}. \tag{2.10}$$

The operator $\mathrm{d}\Gamma(h)$ is called the differential second quantization of h or simply the second quantization of h. Since

$$\mathrm{d}\Gamma(h) = -i\frac{d}{dt}\Gamma(e^{ith})\Big\lceil_{t=0},$$

we have

$$\mathrm{d}\Gamma(h) = h_0 \oplus \left[\oplus_{n=1}^{\infty}\overline{\sum_{j=1}^{n} h_n^j}\right],$$

where $h_0 = 0$, $h_n^j = \underbrace{\mathbb{1}\otimes\cdots\otimes \overset{j}{h}\otimes\cdots\otimes\mathbb{1}}_{n}$ for $n \geq 1$, and the over-line denotes the operator closure. Thus the action of $\mathrm{d}\Gamma(h)$ is given by

$$\mathrm{d}\Gamma(h)\Omega = 0, \quad \mathrm{d}\Gamma(h)a^\dagger(f_1)\cdots a^\dagger(f_n)\Omega = \sum_{j=1}^{n} a^\dagger(f_1)\cdots a^\dagger(hf_j)\cdots a^\dagger(f_n)\Omega$$

for $f_j \in D(h)$, $j = 1, ..., n$. It can be also seen that

---

[5]Let $\{U_t : t \in \mathbb{R}\}$ be a strongly continuous one-parameter unitary group on a Hilbert space $\mathscr{H}$. Then there exists a unique selfadjoint operator A such that $U_t = e^{itA}$ for $t \in \mathbb{R}$. The domain of A is given by $D(A) = \{\psi \in \mathscr{H} \,|\, \lim_{\epsilon\to 0}\frac{1}{\epsilon}(U_\epsilon(\psi) - \psi) \text{ exists}\}$.

$$\mathrm{sp}(\mathrm{d}\Gamma(h)) = \overline{\left\{ \sum_{j=1}^{n} \lambda_j \,\middle|\, \lambda_j \in \mathrm{sp}(h),\ j = 1, ..., n, n \geq 1 \right\}} \cup \{0\},$$

$$\mathrm{sp_p}(\mathrm{d}\Gamma(h)) = \left\{ \sum_{j=1}^{n} \lambda_j \,\middle|\, \lambda_j \in \mathrm{sp_p}(h),\ j = 1, ..., n, n \geq 1 \right\} \cup \{0\}.$$

If $0 \notin \mathrm{sp_p}(h)$, the multiplicity of 0 in $\mathrm{sp_p}(\mathrm{d}\Gamma(h))$ is one. The number operator is defined by the second quantization of the identity operator $\mathbb{1}_{\mathscr{W}}$ on $\mathscr{W}$:

$$N = \mathrm{d}\Gamma(\mathbb{1}_{\mathscr{W}}).$$

Since the action of the number operator is given by

$$N\Omega = 0, \quad Na^\dagger(f_1)\cdots a^\dagger(f_n)\Omega = na^\dagger(f_1)\cdots a^\dagger(f_n)\Omega,$$

the number operator tells how many particles are in a given state and

$$\mathrm{sp}(N) = \mathbb{N} \cup \{0\}.$$

Thus $\mathscr{F}^{(n)}$ is the subspace of eigenvectors of N associated with eigenvalue n. We furthermore extend the second quantization $\mathrm{d}\Gamma(h)$. Let h and k be selfadjoint operators in $\mathscr{W}$. Define $\mathrm{d}\Gamma(k, h)$ by

$$\mathrm{d}\Gamma(k, h) = h_0(k) \oplus \left[\oplus_{n=1}^{\infty} \overline{\sum_{j=1}^{n} h_n^j(k)} \right],$$

where $h_0(k) = 0$ and $h_n^j(k) = \underbrace{k \otimes \cdots \otimes \overset{j}{h} \otimes \cdots \otimes k}_{n}$. From this definition we have $\mathrm{d}\Gamma(h) = \mathrm{d}\Gamma(\mathbb{1}, h)$.

Next we see the relationship between the second quantization and $a^\sharp$. We will use the following facts.

Lemma 2.25 *Let h be a nonnegative selfadjoint operator, $f \in D(h^{-1/2})$ and $\Psi \in D(\mathrm{d}\Gamma(h)^{1/2})$. Then $\Psi \in D(a^\sharp(f))$ and*

$$\|a(f)\Psi\| \leq \|h^{-1/2} f\| \|\mathrm{d}\Gamma(h)^{1/2}\Psi\|,$$
$$\|a^\dagger(f)\Psi\| \leq \|h^{-1/2} f\| \|\mathrm{d}\Gamma(h)^{1/2}\Psi\| + \|f\|\|\Psi\|.$$

In particular, $D(\mathrm{d}\Gamma(h)^{1/2}) \subset D(a^\sharp(f))$, whenever $f \in D(h^{-1/2})$.

To obtain the commutation relations between $a^\sharp(f)$ and $\mathrm{d}\Gamma(h)$, suppose that $f \in D(h^{-1/2}) \cap D(h)$. We have

$$[\mathrm{d}\Gamma(h), a^\dagger(f)]\Psi = a^\dagger(hf)\Psi, \quad [\mathrm{d}\Gamma(h), a(f)]\Psi = -a(\overline{h\bar{f}})\Psi,$$

for $\Psi \in D(\mathrm{d}\Gamma(h)^{3/2}) \cap \mathscr{F}_{\mathrm{fin}}$. We can also see commutation relations:

$$[\mathrm{d}\Gamma(k,h), a^\dagger(f)] = a^\dagger((k-\mathbb{1})f)\mathrm{d}\Gamma(k,h) + a^\dagger(hf)\Gamma(k),$$
$$[\mathrm{d}\Gamma(k,h), a(f)] = -\mathrm{d}\Gamma(k,h)a(\overline{(k-\mathbb{1})\bar{f}}) + \Gamma(k)a(\overline{h\bar{f}}).$$

Let h be a bounded self-adjoint operator. Thus we have

$$\|\mathrm{d}\Gamma(h)\Psi\|^2 = \sum_{n=1}^{\infty} \|\sum_{j=1}^{n} h_n^j \Psi^{(n)}\|^2 \le \sum_{n=1}^{\infty} n^2\|h\|^2\|\Psi^{(n)}\|^2 = \|h\|^2\|N\Psi\|^2.$$

In a similar way we have the lemma below:

Lemma 2.26 *Let h be bounded and $\|k\| \le 1$. Then $\|\mathrm{d}\Gamma(k,h)\Psi\|^2 \le \|h\|^2\| N\Psi\|^2$.*

2.3.3 *Localizations in $\mathscr{F}(\mathscr{W})$*

Later on we shall discuss localization estimates in boson Fock space. Here we introduce an abstract version of localization estimates. Let $j_i : \mathscr{W} \to \mathscr{W}_i$, $i = 1, 2$, be bounded operators such that

$$j_1^* j_1 + j_2^* j_2 = \mathbb{1}_{\mathscr{W}}.$$

We define $j : \mathscr{W} \to \mathscr{W}_1 \oplus \mathscr{W}_2$ by

$$jf = j_1 f \oplus j_2 f.$$

Since $(g_1 \oplus g_2, jf) = (j_1^* g_1 + j_2^* g_2, f)$, it follows that $j^*(g_1 \oplus g_2) = j_1^* g_1 + j_2^* g_2$. Hence $j^* jf = j^*(j_1 f \oplus j_2 f) = (j_1^* j_1 + j_2^* j_2) f = f$, and both j and $\Gamma(j) : \mathscr{F}(\mathscr{W}) \to \mathscr{F}(\mathscr{W}_1 \oplus \mathscr{W}_2)$ are isometries. Let $\bar{U} = \mathscr{U}_c\Gamma(j) : \mathscr{F} \to \mathscr{F}_1 \otimes \mathscr{F}_2$, where $\mathscr{F}_i = \mathscr{F}(\mathscr{W}_i)$. We have the lemma below.

Lemma 2.27 *An intertwining property holds:*

$$\bar{U} a^\sharp(f) = (a^\sharp(j_1 f) \otimes \mathbb{1}_{\mathscr{F}_2} + \mathbb{1}_{\mathscr{F}_1} \otimes a^\sharp(j_2 f))\bar{U}.$$

Let P_Ω be the projection to the subspace of $\mathscr{F}_2$ spanned by the Fock vacuum Ω. We have a useful identity:

Lemma 2.28 *It follows that $\bar{U}^*(\mathbb{1} \otimes P_\Omega)\bar{U} = \Gamma(j_1^* j_1)$.*

Using $\mathrm{d}\Gamma(k,h)$ we can compute

$$\bar{U}\mathrm{d}\Gamma(h) = (\mathrm{d}\Gamma(h) \otimes \mathbb{1} + \mathbb{1} \otimes \mathrm{d}\Gamma(h))\bar{U} - \mathscr{U}_c \mathrm{d}\Gamma(j, (h \oplus h)j - jh).$$

We check that $((h \oplus h)j - jh)f = [h, j_1]f \oplus [h, j_2]f$. Describe $(h \oplus h)j - jh$ as $[h, j]_\oplus$. By Lemma 2.26 if $[h, j]_\oplus$ is bounded with $\|[h, j]_\oplus\| \le C$, we have the bound

$$\|\mathrm{d}\Gamma(j, [h, j]_\oplus)\Psi\| \le C\|N\Psi\|.$$

Thus we have the lemma below:

Lemma 2.29 *It follows that*

$$\bar{U}\mathrm{d}\Gamma(h) = (\mathrm{d}\Gamma(h) \otimes \mathbb{1} + \mathbb{1} \otimes \mathrm{d}\Gamma(h))\bar{U} - \mathscr{U}_\mathrm{c}\mathrm{d}\Gamma(j, [h, j]_\oplus).$$

In particular if $[h, j]_\oplus = 0$, *then* $\mathscr{U}_\mathrm{c}\mathrm{d}\Gamma(j, [h, j]_\oplus)$ *disappears and*

$$\bar{U}\mathrm{d}\Gamma(h) = (\mathrm{d}\Gamma(h) \otimes \mathbb{1} + \mathbb{1} \otimes \mathrm{d}\Gamma(h))\bar{U}.$$

2.4 Boson Fock Space over $L^2(\mathbb{R}^d)$

2.4.1 Field Operators and Second Quantizations

We take now $\mathscr{W} = L^2(\mathbb{R}^d)$ and consider the boson Fock space $\mathscr{F}(L^2(\mathbb{R}^d))$. We have the identification $L^2(\mathbb{R}^{dn}) \cong \otimes^n L^2(\mathbb{R}^d)$ and

$$\prod_{j=1}^n f_j(x_j) \cong (\otimes_{j=1}^n f_j)(x_1, \ldots, x_n).$$

In this case the n-particle subspace $\otimes_\mathrm{s}^n L^2(\mathbb{R}^d)$ can be identified with the set of symmetric functions in $L^2(\mathbb{R}^{dn})$:

$$\otimes_\mathrm{s}^n L^2(\mathbb{R}^d) \cong \{f \in L^2(\mathbb{R}^{dn}) | f(k_1, \ldots, k_n) = f(k_{\pi(1)}, \ldots, k_{\pi(n)}), \forall \pi \in \wp_n\}.$$

We use this identification unless otherwise stated, i.e., for $(\Phi^{(n)})_{n=0}^\infty \in \mathscr{F}(L^2(\mathbb{R}^d))$, $\Phi^{(n)} = \Phi^{(n)}(k_1, \ldots, k_n)$ is a symmetric $L^2(\mathbb{R}^{dn})$-function. The creation operators and the annihilation operators act as

$$(a(f)\Psi)^{(n)}(k_1, ..., k_n) = \sqrt{n+1}\int_{\mathbb{R}^d} f(k)\Psi^{(n+1)}(k, k_1, ..., k_n)dk, \ n \ge 0,$$

$$(a^\dagger(f)\Psi)^{(n)}(k_1, ..., k_n) = \begin{cases} \frac{1}{\sqrt{n}} \sum_{j=1}^n f(k_j)\Psi^{(n-1)}(k_1, ..., \hat{k}_j, ..., k_n), & n \ge 1, \\ 0, & n = 0. \end{cases}$$

Formal description of $a^\sharp(f)$ is given by $a^\sharp(f) = \int a^\sharp(k)f(k)dk$ which is used throughout this book. Let h be a multiplication operator by $h(k)$ in $L^2(\mathbb{R}^d)$. We see that

$$(\Gamma(h)\Psi)^{(n)}(k_1,\ldots,k_n) = \left(\prod_{j=1}^{n} h(k_j)\right) \Psi^{(n)}(k_1,\ldots,k_n),$$

and

$$(\mathrm{d}\Gamma(h)\Psi)^{(n)}(k_1,\ldots,k_n) = \left(\sum_{j=1}^{n} h(k_j)\right) \Psi^{(n)}(k_1,\ldots,k_n).$$

If $0 \le h(k) \le 1$ for all $k \in \mathbb{R}^d$, it follows that

$$0 \le \sum_{n=0}^{\infty}(\Phi^{(n)}, (\mathbb{1} - \prod_{j=1}^{n} h(k_j))\Phi^{(n)}) \le \sum_{n=0}^{\infty}(\Phi^{(n)}, \sum_{j=1}^{n}(\mathbb{1} - h(k_j))\Phi^{(n)}).$$

Thus $\mathbb{1} - \Gamma(h) \le \mathrm{d}\Gamma(\mathbb{1} - h)$ follows in the sense of form. Furthermore it follows that

$$(\mathbb{1} - \Gamma(h))^m \le \mathrm{d}\Gamma(\mathbb{1} - h)$$

for any $m \ge 1$. Let $\hat{h} = h(-i\nabla)$. In a similar manner to this we can also see that

$$(\mathbb{1} - \Gamma(\hat{h}))^m \le \mathrm{d}\Gamma(\mathbb{1} - \hat{h})$$

for any $m \ge 1$.

Now we shall define the free field Hamiltonian in $\mathscr{F}(L^2(\mathbb{R}^d))$. In the special relativity theory an elementary particle with momentum $k \in \mathbb{R}^d$ and rest mass ν has the energy $\sqrt{|k|^2 + \nu^2}$. Let $\omega : L^2(\mathbb{R}^d) \to L^2(\mathbb{R}^d)$ be the multiplication operator by

$$\omega(k) = \sqrt{|k|^2 + \nu^2}$$

with $\nu \ge 0$. Here $\omega(k)$ describes the energy of a single boson with rest mass $\nu \ge 0$ and momentum $k \in \mathbb{R}^d$. ω is called dispersion relation. The selfadjoint operator $\mathrm{d}\Gamma(\omega)$ is called the free field Hamiltonian on $\mathscr{F}(L^2(\mathbb{R}^d))$. We use the notation

$$H_{\mathrm{f}} = \mathrm{d}\Gamma(\omega).$$

Let $\Psi \in \mathscr{F}(L^2(\mathbb{R}^d))$. The free field Hamiltonian H_{f} acts as

$$(H_{\mathrm{f}}\Psi)^{(n)}(k_1, \ldots, k_n) = \left(\sum_{j=1}^{n} \omega(k_j)\right) \Psi^{(n)}(k_1, \ldots, k_n).$$

From this we can see that $H_{\mathrm{f}}\lceil_{\mathscr{F}^{(n)}}$ is the multiplication by $\sum_{j=1}^n \omega(k_j)$. Hence it can be written as $H_{\mathrm{f}} = \{0\} \oplus [\oplus_{n=1}^{\infty} \sum_{j=1}^n \omega(k_j)]$. The spectrum of H_{f} is given by

$\mathrm{sp}(H_\mathrm{f}) = [0, \infty)$ and $\mathrm{sp}_\mathrm{p}(H_\mathrm{f}) = \{0\}$, and 0 is a simple eigenvalue with $H_\mathrm{f}\Omega = 0$. Formally we may write the free field Hamiltonian as $H_\mathrm{f} = \int \omega(k) a^\dagger(k) a(k) dk$. The commutation relations are

$$[H_\mathrm{f}, a(f)] = -a(\omega f), \quad [H_\mathrm{f}, a^\dagger(f)] = a^\dagger(\omega f).$$

The relative bound of $a^\sharp(f)$ with respect to the free field Hamiltonian H_f can be seen as follows. Let $f/\sqrt{\omega} \in L^2(\mathbb{R}^d)$. We have

$$\|a(f)\Psi\| \le \|f/\sqrt{\omega}\| \|H_\mathrm{f}^{1/2}\Psi\|,$$
$$\|a^\dagger(f)\Psi\| \le \|f/\sqrt{\omega}\| \|H_\mathrm{f}^{1/2}\Psi\| + \|f\| \|\Psi\|.$$

Furthermore we can show inequalities below. Suppose that $f_j \in D(1/\sqrt{\omega})$ for $j = 1, \ldots, n$. For $\Phi \in D(H_\mathrm{f}^{n/2})$, we have

$$\left\| \prod_{j=1}^n a(f_j)\Phi \right\| \le \left(\prod_{j=1}^n \|f_j/\sqrt{\omega}\| \right) \|H_\mathrm{f}^{n/2}\Phi\|,$$
$$\left\| \prod_{j=1}^n a^\dagger(f_j)\Phi \right\| \le \sqrt{n!} 2^{n/2} \left(\prod_{l=1}^n \|f_l\|_\omega \right) \left(\sum_{m=0}^n \frac{1}{m!} \|H_\mathrm{f}^{m/2}\Phi\|^2 \right)^{1/2}.$$

Here $\|f\|_\omega = \|f\| + \|f/\sqrt{\omega}\|$. See e.g., [46].

2.4.2 Compact Operators on $\mathscr{F}(L^2(\mathbb{R}^d))$

We consider compact operators on $\mathscr{F}(L^2(\mathbb{R}^d))$ defined by the second quantization of multiplication operators and pseudo differential operators. We have already seen that operators of the form $Q(x)P(-i\nabla)$ are compact on $L^2(\mathbb{R}^d)$ if $P(k)$ and $Q(k)$ decay to zero as $|k| \to \infty$. Similar statements hold for second quantized operators. $\Gamma(Q(k))\Gamma(P(-i\nabla_k))$ leaves the n-particle subspace invariant and the action is

$$\Gamma(Q(k))\Gamma(P(-i\nabla_k)) f(k_1, \ldots, k_n) = \prod_{j=1}^n Q(k_j) \prod_{j=1}^n P(-i\nabla_{k_j}) f(k_1, \ldots, k_n)$$

for $f \in \mathscr{F}^{(n)}$. Hence $\Gamma(Q(k))\Gamma(P(-i\nabla_k))$ restricted on the finite particle subspace is compact.

Proposition 2.30 *Let $F \in C_0^\infty(\mathbb{R}^d)$ and $0 \le F(x) \le 1$ for any $x \in \mathbb{R}^d$. Define $\hat{F} = F(-i\nabla)$. Let P_n be the projection from $\mathscr{F}$ to $\mathscr{F}^{(n)}$. Then $\Gamma(\hat{F})(H_\mathrm{f} + \mathbb{1})^{-1} P_n$ is compact. In particular $\Gamma(\hat{F})\chi(H_\mathrm{f})P_n$ is also compact for any $\chi \in C_0^\infty(\mathbb{R})$.*

Proof Set $K = \Gamma(\hat{F})(H_f + \mathbb{1})^{-1}$. Let $(\Phi_m)_{m=1}^\infty \subset \mathscr{F}$ be a sequence weakly converging to $\Phi \in \mathscr{F}$ as $m \to \infty$. Then $(P_n\Phi_m)_{m=1}^\infty$ also weakly converges to $P_n\Phi$ as $m \to \infty$. Since $P_n\Phi_m \in \mathscr{F}^{(n)} \subset L^2(\mathbb{R}^{dn})$, $(P_n\Phi_m)_{m=1}^\infty$ also weakly converges to $P_n\Phi$ in $L^2(\mathbb{R}^{dn})$ as $m \to \infty$. Operator K acts on $L^2(\mathbb{R}^{dn})$ as

$$Kf(k_1, \ldots, k_n) = \left(\prod_{j=1}^n \hat{F}(-i\nabla_{k_j})\right) \frac{1}{1 + \sum_{j=1}^n \omega(k_j)} f(k_1, \ldots, k_n).$$

By Corollary 2.8, K is compact on $L^2(\mathbb{R}^{dn})$ and hence $KP_n\Phi_m$ strongly converges to $KP_n\Phi$ as $m \to \infty$ in $L^2(\mathbb{R}^{dn})$. Since

$$\|KP_n\Phi - KP_n\Phi_m\|_{\mathscr{F}}^2 = \|KP_n\Phi - KP_n\Phi_m\|_{L^2(\mathbb{R}^{dn})}^2,$$

sequence $(KP_n\Phi_m)_{m=1}^\infty$ also strongly converges to $KP_n\Phi$ as $m \to \infty$ in $\mathscr{F}$. Hence KP_n is compact. Furthermore $\Gamma(\hat{F})\chi(H_f)P_n = KP_n(H_f + \mathbb{1})\chi(H_f)$ and $(H_f + \mathbb{1})\chi(H_f)$ is bounded. Then $\Gamma(\hat{F})\chi(H_f)P_n$ is compact. ■

2.4.3 Localizations in $\mathscr{F}(L^2(\mathbb{R}^d))$

We show localization estimates on $\mathscr{F}(L^2(\mathbb{R}^d))$ which will play an important role throughout this book.

Definition 2.31 We say $(j_1, j_2) \in l_{\mathrm{loc}}(\mathbb{R}^d)$ if and only if

(1) $j_i \in C^\infty(\mathbb{R}^d)$ for $i = 1, 2$,
(2) $j_1^2(k) + j_2^2(k) = 1$ for all $k \in \mathbb{R}^d$,
(3) $0 \le j_i(k) \le 1$ for $i = 1, 2$, and $j_1(k) = \begin{cases} 1, & |k| < 1, \\ 0, & |k| > 2. \end{cases}$

Let $(j_1, j_2) \in l_{\mathrm{loc}}(\mathbb{R}^d)$. We define the multiplication operator $j_{i,R}(k)$ by $j_{i,R}(k) = j_i(k/R)$, where R is a scaling parameter. Note that

$$\lim_{R\to\infty} j_{1,R}(k) = 1 \quad \lim_{R\to\infty} j_{2,R}(k) = 0$$

for each k. Let us define a pseudo differential operator by $\hat{j}_{i,R} = j_i(-i\nabla_k/R)$. Thus we define $\hat{j}_R : L^2(\mathbb{R}^d) \to L^2(\mathbb{R}^d) \oplus L^2(\mathbb{R}^d)$ by $\hat{j}_R f = \hat{j}_{1,R} f \oplus \hat{j}_{2,R} f$. Since $\hat{j}_R$ is isometry, $U_R = \mathscr{U}_c\Gamma(\hat{j}_R) : \mathscr{F}(L^2(\mathbb{R}^d)) \to \mathscr{F}(L^2(\mathbb{R}^d)) \otimes \mathscr{F}(L^2(\mathbb{R}^d))$ is also isometry. More specifically it acts as

$$U_R \prod_{j=1}^n a^\dagger(f_j)\Omega = \prod_{j=1}^n \left(a^\dagger(\hat{j}_{1,R} f_j) \otimes \mathbb{1} + \mathbb{1} \otimes a^\dagger(\hat{j}_{2,R} f_j)\right) \Omega \otimes \Omega.$$

We can also check that

$$U_R^*(\mathbb{1} \otimes P_\Omega)U_R = \Gamma(\hat{j}_{1,R}^* \hat{j}_{1,R}). \tag{2.11}$$

We have

$$U_R^*(\mathbb{1} \otimes P_\Omega)U_R(H_\mathrm{f} + \mathbb{1})^{-1/2} = \left(\prod_{j=1}^n j_1^2(-i\nabla_{k_j}/R)\right)\left(\sum_{j=1}^n \omega(k_j) + 1\right)^{-1/2}$$

on $L^2(\mathbb{R}^{dn})$. Let N be the number operator.

$$U_R^*(\mathbb{1} \otimes P_\Omega)U_R(H_\mathrm{f} + \mathbb{1})^{-1/2}\mathbb{1}_{[0,n]}(N)$$

turns to be a compact operator on $\oplus_{m=0}^n L^2(\mathbb{R}^{dm})$.

Let us consider $H = \phi(f)^{2n} + H_\mathrm{f}$. Let $E = \inf \mathrm{sp}(H)$ and suppose that $F \in D(H)$. We estimate (F, HF) from below. By localization estimate we can have a non-trivial estimate of (F, HF). We divide H into a high energy term and a low energy term by U_R. Since U_R is isometry, we have $(F, HF) = (U_R F, U_R HF)$ and

$$\begin{aligned} U_R H = &\left((a^\dagger(\hat{j}_{1,R}f) + a(\hat{j}_{1,R}f)) \otimes \mathbb{1} + \mathbb{1} \otimes (a^\dagger(\hat{j}_{2,R}f)) + a(\hat{j}_{2,R}f)\right)^{2n} U_R \\ &+(H_\mathrm{f} \otimes \mathbb{1} + \mathbb{1} \otimes H_\mathrm{f})U_R + \mathscr{U}_\mathrm{c}\mathrm{d}\Gamma(\hat{j}_R, [\omega, \hat{j}_R]_\oplus). \end{aligned}$$

Roughly speaking we see that the high energy term $a^\dagger(\hat{j}_{2,R}f)) + a(\hat{j}_{2,R}f)$ reduces to zero as $R \to \infty$, and $\mathrm{d}\Gamma(\hat{j}_R, [\omega, \hat{j}_R])$ also reduces to zero. Furthermore $\hat{j}_{1,R}f \to f$ as $R \to \infty$ implies that

$$(F, HF) \sim (U_R F, \left\{(a^\dagger(f) + a(f))^{2n} + H_\mathrm{f}\right\} \otimes \mathbb{1}U_R F) + (U_R F, (\mathbb{1} \otimes H_\mathrm{f})U_R F) + o(R)$$

as $R \to \infty$. The low energy term has a trivial bound

$$(U_R F, \left\{(a^\dagger(f) + a(f))^{2n} + H_\mathrm{f}\right\} \otimes \mathbb{1}U_R F) \geq E\|F\|^2,$$

and the high energy term by

$$(U_R F, (\mathbb{1} \otimes H_\mathrm{f})U_R F) \geq \nu\|F\|^2 - \nu(U_R F, (P_\Omega \otimes \mathbb{1})U_R F),$$

where we used $H_\mathrm{f} \geq \nu(\mathbb{1} - P_\Omega)$. Thus we obtain by (2.11)

$$(F, HF) \geq (E + \nu)\|F\|^2 - \nu(F, \Gamma(\hat{j}_{1,R}^* \hat{j}_{1,R})F).$$

This type of argument will be used often times in this book.

Chapter 3
The Pauli–Fierz Model

Abstract In this chapter the Pauli–Fierz model in non-relativistic quantum electrodynamics is studied. This model describes the minimal interaction between quantum matters (electrons) and a massless quantized radiation field (photons). The existence of the ground state of the Pauli–Fierz Hamiltonian is proven.

3.1 The Pauli–Fierz Hamiltonian

We consider a system of quantum matters minimally coupled to a quantized radiation field. This model describes an interaction between non-relativistic spinless N-electrons and photons. Suppose that the space dimension is 3 and the number of electron is one for simplicity. Let

$$\mathscr{H} = L^2(\mathbb{R}^3) \otimes \mathscr{F}$$

be the total Hilbert space describing the electron-photon state vectors. $L^2(\mathbb{R}^3)$ describes the state space of a single electron moving in $\mathbb{R}^3$ and $\mathscr{F}$ that of photons. Here $\mathscr{F} = \mathscr{F}(\mathscr{H}_{\mathrm{PF}})$ is the boson Fock space over Hilbert space

$$\mathscr{H}_{\mathrm{PF}} = L^2(\mathbb{R}^3 \times \{1, 2\})$$

of the set of L^2-functions on $\mathbb{R}^3 \times \{1, 2\}$. The elements of the set $\{1, 2\}$ account for the fact that a photon is a transversal wave perpendicular to the direction of its propagation, which has two components. $\mathscr{H}$ can be decomposed into infinite direct sum:

$$\mathscr{H} = \oplus_{n=0}^{\infty} \mathscr{H}^{(n)},$$

where $\mathscr{H}^{(n)} = L^2(\mathbb{R}^3) \otimes \mathscr{F}^{(n)}$. The Fock vacuum in $\mathscr{F}$ is denoted by Ω as usual. We introduce the free field Hamiltonian on $\mathscr{F}$. Let $\omega = \omega(k) = |k|$ which describes the energy of a single photon with momentum k. Note that a photon is massless. The free field Hamiltonian H_{rad} on $\mathscr{F}$ is given in terms of the second quantization

F. Hiroshima, *Ground States of Quantum Field Models*,
SpringerBriefs in Mathematical Physics,
https://doi.org/10.1007/978-981-32-9305-2_3

$$H_{\rm rad} = {\rm d}\Gamma(\omega).$$

Here ω is regarded as the multiplication in $\mathscr{H}_{\rm PF}$ by $(\omega f)(k, j) = \omega(k) f(k, j)$ for $(k, j) \in \mathbb{R}^3 \times \{1, 2\}$.

On the other hand the quantum matter, electron, is governed by a Schrödinger operator of the form

$$H_{\rm p} = -\frac{1}{2m}\Delta + V$$

in $L^2(\mathbb{R}^3)$. Here m denotes the mass of an electron. To introduce the minimal coupling we define quantized radiation fields. Let $a(f)$ and $a^\dagger(f)$ be the annihilation operator and the creation operator on $\mathscr{F}$ smeared by $f \in \mathscr{H}_{\rm PF}$, respectively. Let us identify $\mathscr{H}_{\rm PF}$ with $L^2(\mathbb{R}^3) \oplus L^2(\mathbb{R}^3)$ by

$$\mathscr{H}_{\rm PF} \ni f(\cdot, 1) \cong f(\cdot, 1) \oplus 0 \in L^2(\mathbb{R}^3) \oplus L^2(\mathbb{R}^3),$$
$$\mathscr{H}_{\rm PF} \ni f(\cdot, 2) \cong 0 \oplus f(\cdot, 2) \in L^2(\mathbb{R}^3) \oplus L^2(\mathbb{R}^3).$$

We set $a^\sharp(f \oplus 0) = a^\sharp(f, 1)$ and $a^\sharp(0 \oplus f) = a^\sharp(f, 2)$. Hence we obtain canonical commutation relations:

$$[a(f, j), a^\dagger(g, j')] = \delta_{jj'}(\bar{f}, g), \quad [a(f, j), a(g, j')] = 0 = [a^\dagger(f, j), a^\dagger(g, j')].$$

We define the quantized radiation field with a cutoff function $\hat{\varphi}$. Put

$$\wp_\mu(x, j) = \frac{\hat{\varphi}(k)}{\sqrt{\omega(k)}} e_\mu(k, j) e^{-ikx}, \quad \tilde{\wp}_\mu(x, j) = \frac{\hat{\varphi}(-k)}{\sqrt{\omega(k)}} e_\mu(k, j) e^{ikx}$$

for each $x \in \mathbb{R}^3$, $j = 1, 2$ and $\mu = 1, 2, 3$. Here cutoff function $\hat{\varphi}$ is the Fourier transform of the charge distribution $\varphi \in \mathscr{S}'(\mathbb{R}^3)$. Although physically it should be $\hat{\varphi} = 1/(2\pi)^{3/2}$, we have to introduce cutoff function $\hat{\varphi}$ to ensure that $\wp_\mu(x, j) \in L^2(\mathbb{R}^3_k)$ for each x and j. The vectors $e(k, 1)$ and $e(k, 2)$ are called polarization vectors, that is, $(e(k, 1), e(k, 2), k/|k|)$ forms a right-hand system at each $k \in \mathbb{R}^3 \setminus \{0\}$;

$$e(k, i) \cdot e(k, j) = \delta_{ij}, \quad e(k, j) \cdot k = 0, \quad e(k, 1) \times e(k, 2) = \frac{k}{|k|}.$$

The quantized radiation field with cutoff function $\hat{\varphi}$ is defined by

$$A_\mu(x) = \frac{1}{\sqrt{2}} \sum_{j=1,2} (a^\dagger(\wp_\mu(x, j), j) + a(\tilde{\wp}_\mu(x, j), j)), \quad \mu = 1, 2, 3.$$

Unless otherwise stated we suppose the following assumptions.

Assumption 3.1 (*Cutoff functions*) $\varphi \in \mathscr{S}'(\mathbb{R}^3)$ satisfies that (1) $\hat{\varphi} \in L^1_{\rm loc}(\mathbb{R}^3)$, (2) $\hat{\varphi}(-k) = \overline{\hat{\varphi}(k)}$, (3) $\sqrt{\omega}\hat{\varphi}, \hat{\varphi}/\sqrt{\omega}, \hat{\varphi}/\omega \in L^2(\mathbb{R}^3)$.

By $\hat{\varphi}/\sqrt{\omega} \in L^2(\mathbb{R}^3)$ and $\overline{\hat{\varphi}(k)} = \hat{\varphi}(-k)$, for each x, $A_\mu(x)$ is symmetric, and moreover essentially selfadjoint on the finite particle subspace $\mathscr{F}_{\mathrm{fin}}$ of $\mathscr{F}$. We denote the closure of $A_\mu(x)\lceil_{\mathscr{F}_{\mathrm{fin}}}$ by the same symbol. Write

$$A_\mu = \int_{\mathbb{R}^3}^{\oplus} A_\mu(x)dx, \quad A = (A_1, A_2, A_3).$$

A_μ is a selfadjoint operator on

$$D(A_\mu) = \left\{ F \in \mathscr{H} \,\middle|\, F(x) \in D(A_\mu(x)) \text{ a.e. and } \int_{\mathbb{R}^3} \|A_\mu(x)F(x)\|^2_{\mathscr{F}} dx < \infty \right\}$$

and acts as $(A_\mu F)(x) = A_\mu(x)F(x)$ for $F \in D(A_\mu)$ for a.e. $x \in \mathbb{R}^3$. In terms of the formal kernel $a^\sharp(k, j)$, $a^\sharp(f, j)$ can be written as $a^\sharp(f) = \sum_{j=1,2} \int a^\sharp(k, j)dk$, and the quantized radiation field $A_\mu(x)$ as

$$A_\mu(x) = \sum_{j=1,2} \frac{1}{\sqrt{2}} \int e_\mu(k, j) \left(\frac{\hat{\varphi}(k)}{\sqrt{\omega(k)}} e^{-ikx} a^\dagger(k, j) + \frac{\hat{\varphi}(-k)}{\sqrt{\omega(k)}} e^{ikx} a(k, j) \right) dk.$$

Since $k \cdot e(k, j) = 0$, the polarization vectors introduced above are chosen in the way that $\sum_{\mu=1}^{3} \nabla_\mu \wp_j^\mu(x) = 0$, implying the Coulomb gauge condition:

$$\sum_{\mu=1}^{3} \nabla_\mu A_\mu = 0.$$

This in turn yields $\sum_{\mu=1}^{3} [\nabla_\mu, A_\mu] = 0$.

Let us define the Pauli–Fierz Hamiltonian. The interaction is obtained by the minimal coupling: $-i\nabla_\mu \otimes \mathbb{1} \mapsto -i\nabla_\mu \otimes \mathbb{1} - \alpha A_\mu$ to $H_{\mathrm{p}} \otimes \mathbb{1} + \mathbb{1} \otimes H_{\mathrm{rad}}$, where α denotes a coupling constant.

Definition 3.2 (*The Pauli–Fierz Hamiltonian*) The Pauli–Fierz Hamiltonian of one electron with mass m is defined by

$$H_{\mathrm{PF}} = \frac{1}{2m}(-i\nabla \otimes \mathbb{1} - \alpha A)^2 + V \otimes \mathbb{1} + \mathbb{1} \otimes H_{\mathrm{rad}}.$$

In what follows we set $m = 1$, $\alpha = 1$ and omit the tensor notation $\otimes$ for the sake of simplicity. Thus

$$H_{\mathrm{PF}} = \frac{1}{2}(-i\nabla - A)^2 + V + H_{\mathrm{rad}}.$$

We introduce classes of external potentials.

Definition 3.3 (V_{Kato}) We say $V \in V_{\mathrm{Kato}}$ if and only if $D(\Delta) \subset D(V)$ and there exist $0 \le a < 1$ and $0 \le b$ such that $\|Vf\| \le a\| - (1/2)\Delta f\| + b\|f\|$ for $f \in D(\Delta)$.

Proposition 3.4 (Selfadjointness) *Suppose $V \in V_{\rm Kato}$. Then $H_{\rm PF}$ is selfadjoint and bounded from below on $D(H_{\rm p}) \cap D(H_{\rm rad})$ and essentially selfadjoint on any core of $H_{\rm p} + H_{\rm rad}$.*

Proof See e.g. [15, 23, 32, 34, 45]. ■

Note that the Pauli–Fierz Hamiltonians with different polarization vectors are equivalent with each other. We can show this below. Let e^1, e^2 and η^1, η^2 be polarization vectors, and $H_{\rm PF}(e^1, e^2)$ and $H_{\rm PF}(\eta^1, \eta^2)$ the corresponding Pauli–Fierz Hamiltonians, respectively. In Proposition 3.5 we will see that the spectral analysis is independent of the choice of polarization vectors, i.e., the Hamiltonians defined through different set of polarizations are unitary equivalent. Thus we may fix polarization vectors as it is most convenient.

Proposition 3.5 *Let (e^1, e^2) and (η^1, η^2) be polarization vectors. Then $H_{\rm PF}(e^1, e^2)$ and $H_{\rm PF}(\eta^1, \eta^2)$ are unitary equivalent.*

Proof Since for each $k \in \mathbb{R}^3$ both polarization vectors form orthogonal bases on the plane perpendicular to the vector k, there exists $\theta_k \in [0, 2\pi)$ such that

$$\begin{pmatrix} e^1(k) \\ e^2(k) \end{pmatrix} = \begin{pmatrix} \cos\theta_k \mathbb{1}_3 & -\sin\theta_k \mathbb{1}_3 \\ \sin\theta_k \mathbb{1}_3 & \cos\theta_k \mathbb{1}_3 \end{pmatrix} \begin{pmatrix} \eta^1(k) \\ \eta^2(k) \end{pmatrix} \quad \text{i.e.} \quad \begin{pmatrix} e^1_\mu(k) \\ e^2_\mu(k) \end{pmatrix} = R_k \begin{pmatrix} \eta^1_\mu(k) \\ \eta^2_\mu(k) \end{pmatrix},$$

where $R_k = \begin{pmatrix} \cos\theta_k & -\sin\theta_k \\ \sin\theta_k & \cos\theta_k \end{pmatrix}$ and $\mathbb{1}_3$ denotes 3-dimensional identity. Define the unitary $R : L^2(\mathbb{R}^3; \mathbb{C}^2) \to L^2(\mathbb{R}^3; \mathbb{C}^2)$ by $R\begin{pmatrix} f \\ g \end{pmatrix}(k) = R_k \begin{pmatrix} f(k) \\ g(k) \end{pmatrix}$ a.e. and $\Gamma(R) : \mathscr{F} \to \mathscr{F}$ by the second quantization of R. $\Gamma(R)$ is also unitary on $\mathscr{F}$, and $\Gamma(R) H_{\rm PF}(\eta^1, \eta^2) \Gamma(R)^{-1} = H_{\rm PF}(e^1, e^2)$. Hence the proof is complete. ■

We fix polarization vectors through out this chapter:

$$e(k, 1) = \frac{(-k_2, k_1, 0)}{\sqrt{k_1^2 + k_2^2}}, \quad e(k, 2) = \frac{(k_3 k_1, -k_2 k_3, k_1^2 + k_2^2)}{|k|\sqrt{k_1^2 + k_2^2}}. \tag{3.1}$$

We can also define the Pauli–Fierz Hamiltonian in more general settings to treat more general external potentials. For the sake of simplicity we set $P_A = -i\nabla - A$. Hence $H_{\rm PF} = \frac{1}{2} P_A^2 + H_{\rm rad} + V$ and $H_{\rm PF}$ defines the quadratic form q on $\mathscr{H}$ by

$$\mathrm{q}(F, G) = \frac{1}{2}(P_A F, P_A G) + (H_{\rm rad}^{1/2} F, H_{\rm rad}^{1/2} G) + (V_+^{1/2} F, V_+^{1/2} G) - (V_-^{1/2} F, V_-^{1/2} G)$$

and

$$\mathrm{q}_0(F, G) = \frac{1}{2}(P_A F, P_A G) + (H_{\rm rad}^{1/2} F, H_{\rm rad}^{1/2} G).$$

Here $V_+(x) = \max\{V(x), 0\}$ and $V_-(x) = -\min\{V(x), 0\}$. The form domain of q is $Q(\mathrm{q}) = D(\sqrt{-\Delta}) \cap D(V_+^{1/2}) \cap D(V_-^{1/2}) \cap D(H_{\mathrm{rad}}^{1/2})$. We introduce a class of external potentials.

Definition 3.6 ($V_{\mathrm{Kato}}^{\mathrm{q}}$) We say that $V = V_+ - V_- \in V_{\mathrm{Kato}}^{\mathrm{q}}$ if and only if $Q(\mathrm{q})$ is dense, $V_+^{1/2} \in L^2_{\mathrm{loc}}(\mathbb{R}^3)$, $D(\sqrt{-\Delta}) \subset D(V_-^{1/2})$ and for any $\varepsilon > 0$ there exists $b_\varepsilon \geq 0$ such that

$$\|V_-^{1/2} f\| \leq \varepsilon \|\sqrt{-\Delta} f\| + b_\varepsilon \|f\|, \quad f \in D(\sqrt{-\Delta}).$$

Lemma 3.7 *Suppose that $V \in V_{\mathrm{Kato}}^{\mathrm{q}}$. Then q is a symmetric, closed and semi-bounded quadratic form. Furthermore $C_0^\infty(\mathbb{R}^3) \otimes \left(\mathscr{F} \cap D(H_{\mathrm{rad}}^{1/2})\right)$ is a form core of q_o.*

Proof It is trivial to see that q is symmetric. For all $\varepsilon > 0$ there exist constants $C_1 > 0$ and $C_2 > 0$ such that for all $F \in Q(\mathrm{q})$

$$\frac{1}{2}((-i\nabla - A)F, (-i\nabla - A)F) + \varepsilon \|H_{\mathrm{rad}}^{1/2} F\|^2 \geq C_1 \|\sqrt{-\Delta} F\|^2 - C_2 \|F\|^2.$$

From this inequality we can have

$$\mathrm{q}(F, F) \geq C_1 \|\sqrt{-\Delta} F\|^2 - C_2 \|F\|^2 + \|V_+^{1/2} F\|^2 - \|V_-^{1/2} F\|^2$$

and

$$\mathrm{q}(F, F) \geq C_1' \|\sqrt{-\Delta} F\|^2 - C_2' \|F\|^2 + \|V_+^{1/2} F\|^2 \geq -C_2' \|F\|^2. \tag{3.2}$$

Then q is semi-bounded. Suppose $\mathrm{q}(F_n - F_m, F_n - F_m) \to 0$ and $\|F_n - F_m\| \to 0$ as $n, m \to \infty$. From (3.2) $H_{\mathrm{rad}}^{1/2} F_n$, $V_+ F_n$ and $\sqrt{-\Delta} F_n$ are Cauchy sequences. There exists $F \in D(H_{\mathrm{rad}}^{1/2}) \cap D(\sqrt{-\Delta}) \cap D(V_+^{1/2})$ such that $\mathrm{q}(F_n - F, F_n - F) \to 0$ as $n \to \infty$. Hence q is closed. Let $V = 0$. We also have the inequality

$$\mathrm{q}_o(F, F) \leq C(\|\sqrt{-\Delta} F\|^2 + \|H_{\mathrm{rad}}^{1/2} F\|^2 + \|F\|^2).$$

Since $C_0^\infty(\mathbb{R}^3) \otimes \left(\mathscr{F} \cap D(H_{\mathrm{rad}}^{1/2})\right)$ is a form core of the form on the right-hand side above, $C_0^\infty(\mathbb{R}^3) \otimes \left(\mathscr{F} \cap D(H_{\mathrm{rad}}^{1/2})\right)$ is also a form core of q_o. ■

Let $E = \inf_{F \in Q(\mathrm{q}), \|F\|=1} \mathrm{q}(F, F)$. By Lemma 3.7 and the second representation theorem for quadratic forms [40, p. 331] there exists a unique selfadjoint operator $H_{\mathrm{PF}}^{\mathrm{q}}$ in $\mathscr{H}$ such that $\inf \mathrm{sp}(H_{\mathrm{PF}}^{\mathrm{q}}) = E$, $Q(\mathrm{q}) = D(|H_{\mathrm{PF}}^{\mathrm{q}}|^{1/2})$ and

$$\mathrm{q}(F, G) - E(F, G) = ((H_{\mathrm{PF}}^{\mathrm{q}} - E)^{1/2} F, (H_{\mathrm{PF}}^{\mathrm{q}} - E)^{1/2} G)$$

for all $F, G \in Q(\mathrm{q})$. We define the scalar product on $Q(q)$ by

$$(F, G)_+ = \mathrm{q}(F, G) - E(F, G).$$

Then $(Q(\mathrm{q}), (\cdot, \cdot)_+)$ becomes a Hilbert space.

If $V \in V_{\mathrm{Kato}}$, the Pauli–Fierz Hamiltonian is given by H_{PF}, and if $V \in V_{\mathrm{Kato}}^{\mathrm{q}}$, it is given by $H_{\mathrm{PF}}^{\mathrm{q}}$. In what follows we write H_{PF} for $H_{\mathrm{PF}}^{\mathrm{q}}$.

3.2 Minimizer of Quadratic Forms

We show a general lemma concerning essential spectrum of selfadjoint operators. Following proposition is well known.

Proposition 3.8 (Weyl's criterion) *Let K be a selfadjoint operator in a Hilbert space $\mathscr{K}$. Then $z \in \mathrm{sp}_{\mathrm{ess}}(K)$ if and only if there exists a sequence $\{f_n\}_{n=1}^{\infty} \subset D(K)$ such that $\|f_n\| = 1$, $w - \lim_{n\to\infty} f_n = 0$ and $\lim_{n\to\infty} \|(K - z) f_n\| = 0$.*

Proof See e.g., [1, Proposition 4.20]. ■

The sequence $\{f_n\}_{n=1}^{\infty}$ in Proposition 3.8 is called a Weyl sequence.

Now let us state an abstract proposition. Let r be a closed, symmetric quadratic form on a Hilbert space $\mathscr{K}$ bounded below. Then there exists a unique selfadjoint operator K such that $\mathrm{r}(f, g) - E(f, g) = ((K - E)^{1/2} f, (K - E)^{1/2} g)$, where

$$E = \inf_{f \in Q(\mathrm{r})} \frac{\mathrm{r}(f, f)}{\|f\|^2} \tag{3.3}$$

by the second representation theorem for quadratic forms.

Proposition 3.9 *Let E be* (3.3). *Suppose that* $\liminf_{n\in\mathbb{N}} \mathrm{r}(f_n, f_n) > E$ *whenever $\|f_n\| = 1$ and $w - \lim_{n\to\infty} f_n = 0$. Then E belongs to the discrete spectrum of K, i.e., E is an eigenvalue of K.*

Proof We prove this by contraction. Suppose that E lies in the essential spectrum of K. Then 0 belongs to the essential spectrum of $(K - E)^{1/2}$ and there exists a Weyl sequence $\{f_n\}_{n=1}^{\infty} \subset D(K)$ such that $\|f_n\| = 1$, $w - \lim_{n\to\infty} f_n = 0$ and $\lim_{n\to\infty} \|(K - E)^{1/2} f_n\| = 0$. It follows however that

$$0 = \lim_{n\to\infty} \|(K - E)^{1/2} f_n\|^2 = \lim_{n\to\infty} (\mathrm{r}(f_n, f_n) - E) = \liminf_{n\to\infty} (\mathrm{r}(f_n, f_n) - E) > 0.$$

It contradicts. Hence 0 belongs to the discrete spectrum of $(K - E)^{1/2}$ and hence E also belongs to the discrete spectrum of K. ■

An alternative proof of Proposition 3.9 [21, Theorem 4.1]:

Proof Let $\{F_n\}_{n=1}^\infty \subset Q(\mathrm{q})$ be a minimizing sequence of q. i.e., $\|F_n\| = 1$ and $\lim_{n\to\infty} q(F_n, F_n) = E$. Since $\mathrm{q}(F_n, F_n) \to E$ as $n \to \infty$ and $\|F_n\| = 1$, $\{\|F_n\|_+\}_{n=1}^\infty$ is a bounded sequence. Since $(Q(\mathrm{q}), (\cdot,\cdot)_+)$ is a Hilbert space, taking a subsequence n_k, $\{F_{n_k}\}_{k=1}^\infty$ weakly converges to some $F \in Q(\mathrm{q})$ in $(Q(\mathrm{q}), (\cdot,\cdot)_+)$. In particular $\mathrm{q}(G, F_{n_k}) \to \mathrm{q}(G, F)$ as $n_k \to \infty$ for any $G \in Q(\mathrm{q})$. We replace n_k with n. Putting $\tilde{F}_n = F - F_n$ we obtain that

$$0 = \lim_{n\to\infty}(\mathrm{q}(F_n, F_n) - E) = \lim_{n\to\infty}(\mathrm{q}(\tilde{F}_n, \tilde{F}_n) - E(\tilde{F}_n, \tilde{F}_n)) + \mathrm{q}(F, F) - E(F, F). \tag{3.4}$$

Since $\mathrm{q}(\tilde{F}_n, \tilde{F}_n) - E(\tilde{F}_n, \tilde{F}_n) \geq 0$ and $\mathrm{q}(F, F) - E(F, F) \geq 0$, together with (3.4) we conclude that $\lim_{n\to\infty}(\mathrm{q}(\tilde{F}_n, \tilde{F}_n) - E(\tilde{F}_n, \tilde{F}_n)) = 0$ and $\mathrm{q}(F, F) = E(F, F)$. Thus it is sufficient to show that $F \neq 0$ for the the existence of a minimizer. If $F = 0$, $\|\tilde{F}_n\|^2 = 1$ for any n. It suffices to show that $\xi = \inf_{n\in\mathrm{N}} \|\tilde{F}_n\| = 0$. Suppose that $\xi > 0$. For any $G \in \mathscr{H}$ we have

$$|(G, \frac{\tilde{F}_n}{\|\tilde{F}_n\|})| \leq \frac{|(G, \tilde{F}_n)|}{\xi} \to 0$$

as $n \to \infty$. It implies that $\tilde{F}_n/\|\tilde{F}_n\|$ weakly converges to zero, hence by the assumption we see that

$$0 < \liminf_{n\to\infty} \mathrm{q}(\frac{\tilde{F}_n}{\|\tilde{F}_n\|}, \frac{\tilde{F}_n}{\|\tilde{F}_n\|}) - E \leq \lim_{n\to\infty} \frac{\mathrm{q}(\tilde{F}_n, \tilde{F}_n) - E(\tilde{F}_n, \tilde{F}_n)}{\xi^2} = 0.$$

This contradicts. Hence $\xi = 0$ and the proposition is proven. ■

3.3 Existence of the Ground State for Massive Case $\nu > 0$

In this section we prove the existence of the ground state of the Pauli–Fierz Hamiltonian with an artificial positive mass. Throughout this section we limit ourselves to considering H_{PF} with a dispersion relation which possesses an artificial positive mass $\nu > 0$:

$$\omega(k) = \sqrt{|k|^2 + \nu^2}, \quad \nu > 0.$$

First of all we define ionization energy of H_{PF}. For $(\chi, \tilde{\chi}) \in l_{\mathrm{loc}}(\mathbb{R}^3)$ we define $\chi_R(x) = \chi(x/R)$ and $\tilde{\chi}_R(x) = \tilde{\chi}(x/R)$ with parameter $R > 0$. Thus we define E_R by

$$E_R = \inf_{F\in\mathcal{Q}(\mathrm{q}),\tilde\chi_R F\neq 0} \frac{\mathrm{q}(\tilde\chi_R F, \tilde\chi_R F)}{(\tilde\chi_R F, \tilde\chi_R F)}. \tag{3.5}$$

Here notice that $\tilde\chi_R F \in \mathcal{Q}(\mathrm{q})$. Let

$$\Sigma = \lim_{R\to\infty} E_R.$$

E_R is finite since q is bounded from below, but Σ may take value $+\infty$. Let

$$E_0 = \inf_{F\in\mathcal{Q}(\mathrm{q}_0)} \frac{\mathrm{q}_0(F, F)}{\|F\|^2}.$$

Assumption 3.10 (*Binding condition*) We assume that $E < \Sigma$ for $\nu \geq 0$.

If $\lim_{|x|\to\infty} V(x) = \infty$, the binding condition holds. We show a non-trivial example below.

Proposition 3.11 *Suppose that $V(x) \leq 0$, $\lim_{|x|\to\infty} V(x) = 0$ and for sufficiently large $|x|$, $V(x) \leq -a/|x|^\alpha$ for $0 < \alpha < 2$ and $a > 0$. Then the binding condition holds.*

Proof Let $(\chi, \tilde\chi) \in l_{\mathrm{loc}}(\mathbb{R}^3)$, and $\chi_R(x) = \chi(x/R)$ and $\tilde\chi_R(x) = \tilde\chi(x/R)$. Using the identity $\chi_R\partial_\mu\chi_R + \tilde\chi_R\partial_\mu\tilde\chi_R = 0$ for $\mu = 1, 2, 3$, we have IMS localization formula

$$\mathrm{q}(F, F) = \mathrm{q}(\tilde\chi_R F, \tilde\chi_R F) + \mathrm{q}(\chi_R F, \chi_R F) - \frac{1}{2}\||\nabla\chi_R|F\|^2 - \frac{1}{2}\||\nabla\tilde\chi_R|F\|^2.$$

Set $\tilde\chi_R F = F_{\bar R}$ and $\chi_R F = F_R$. We have

$$\begin{aligned}
&\mathrm{q}(F, F) - \mathrm{q}_0(F_{\bar R}, F_{\bar R}) \leq \mathrm{q}(F, F) - \mathrm{q}(F_{\bar R}, F_{\bar R}) \leq \mathrm{q}(F_R, F_R)\\
&\leq \mathrm{q}(F, F) - \mathrm{q}_0(F_{\bar R}, F_{\bar R}) + \frac{C}{R^2} = \mathrm{q}_0(F, F) - \mathrm{q}_0(F_{\bar R}, F_{\bar R}) + \frac{C}{R^2} - \|V_-^{1/2} F_R\|^2.
\end{aligned}$$

Then

$$\mathrm{q}(F, F) \leq \mathrm{q}_0(F, F) + \frac{C}{R^2} - \|V_-^{1/2} F_R\|^2 \leq \mathrm{q}_0(F, F) + \frac{C}{R^2} - \frac{a}{R^\alpha}.$$

Let $\varepsilon > 0$. Take F such that $\mathrm{q}_0(F, F) \leq E_0 + \varepsilon$.

$$E \leq \mathrm{q}(F, F) \leq E_0 + \varepsilon + \frac{C}{R^2} - \frac{a}{R^\alpha} < E_0$$

follows for ε and R such that $\varepsilon + \frac{C}{R^2} - \frac{a}{R^\alpha} < 0$. Then $E < E_0$. Let

$$E_{0,R} = \inf_{F\in\mathcal{Q}(\mathrm{q}_0),\tilde\chi_R F\neq 0} \frac{\mathrm{q}_0(\tilde\chi_R F, \tilde\chi_R F)}{(\tilde\chi_R F, \tilde\chi_R F)}.$$

We shall prove that

$$\lim_{R\to\infty} E_R = \lim_{R\to\infty} E_{0,R}. \tag{3.6}$$

If this is true, we see that $\lim_{R\to\infty} E_R = \lim_{R\to\infty} E_{0,R} \geq E_0 > E$ and the binding condition is satisfied. It is trivial to see that $\lim_{R\to\infty} E_R \leq \lim_{R\to\infty} E_{0,R}$. We show the inequality in the opposite direction. By the definition of E_R there exists a normalized F such that

$$E_R + \frac{\varepsilon}{2} \geq \mathrm{q}_0(F_{\bar{R}}, F_{\bar{R}}) - \frac{a}{R^\alpha}$$

and

$$\mathrm{q}_0(F_{\bar{R}}, F_{\bar{R}}) - \frac{a}{R^\alpha} \geq E_{0,R} - \frac{\varepsilon}{2}$$

for sufficiently large R. Thus $E_R + \varepsilon \geq E_{0,R}$ for sufficiently large R and

$$\lim_{R\to\infty} E_R + \varepsilon \geq \lim_{R\to\infty} E_{0,R}.$$

Then (3.6) is proven. ■

Corollary 3.12 *Suppose that* $V(x) \leq 0$, $\lim_{|x|\to\infty} V(x) = 0$ *and for sufficiently large* $|x|$, $V(x) \leq -a/|x|^\alpha$ *for* $0 < \alpha < 2$ *and* $a > 0$. *Then* $E < E_0 \leq \Sigma$ *holds.*

Proof In the proof of Proposition 3.11 $E < E_0$ is shown. Since $E_0 \leq \lim_{R\to\infty} E_{0,R}$ and $\lim_{R\to\infty} E_{0,R} = \lim_{R\to\infty} E_R = \Sigma$, the corollary follows. ■

Remark 3.13 In general it can be proven that $\Sigma = E_0$ and more general identities concerning the N-body Pauli–Fierz Hamiltonian are shown in [20, Theorem 3].

The outline of the strategy to show the existence of the ground state is as follows. Suppose that a sequence of normalized vectors $\{F_n\}_{n=1}^\infty$ weakly converges to 0 as $n \to \infty$. Roughly speaking $\| |\nabla \chi_R| F_n\|^2 + \| |\nabla \tilde{\chi}_R| F_n\|^2 \sim 1/R^2$. Hence as was seen in Sect. 2.4.3 we expect that

$$\begin{aligned} &\mathrm{q}(\chi_R F_n, \chi_R F_n) \geq (E + \nu)\|\chi_R F_n\|^2 + \|KLF_n\|^2, \\ &\mathrm{q}(\tilde{\chi}_R F_n, \tilde{\chi}_R F_n) \geq E_R \|\tilde{\chi}_R F_n\|^2 \end{aligned}$$

by localization estimates in boson Fock space. Here K is a compact operator and L bounded. This yields that

$$\begin{aligned} \mathrm{q}(F_n, F_n) &\geq E + \nu\|\chi_R F_n\|^2 + (E_R - E)\|\tilde{\chi}_R F_n\|^2 + \|KLF_n\|^2 \\ &\geq E + \min\{\nu, \Sigma - E\} + \|KLF_n\|^2. \end{aligned}$$

Since K is compact and $\nu > 0$ and $\Sigma - E > 0$ are assumed, we obtain that

$$\liminf_{n\to\infty} \mathrm{q}(F_n, F_n) > E.$$

This implies that H_{PF} admits a ground state F by Proposition 3.9. This can be rigorously proven.

To estimate $\mathrm{q}(\tilde{\chi}_R F, \tilde{\chi}_R F)$ we proceed a localization estimate. Let $(j_1, j_2) \in l_{\mathrm{loc}}(\mathbb{R}^3)$. We define the multiplication operator $j_{i,P}(k)$ by $j_{i,P}(k) = j_i(k/P)$ and the pseudo differential operator $\hat{j}_{i,P}$ by $\hat{j}_{i,P} = j_i(-i\nabla_k/P)$. We define $\hat{j}_P : \mathscr{H}_{\mathrm{PF}} \to \mathscr{H}_{\mathrm{PF}} \oplus \mathscr{H}_{\mathrm{PF}}$ by

$$\hat{j}_P = \hat{j}_{1,P} \oplus \hat{j}_{2,P}.$$

$\hat{j}_P$ is isometry and $U_P = \mathscr{U}_{\mathrm{c}} \Gamma(\hat{j}_P) : \mathscr{F}(\mathscr{H}_{\mathrm{PF}}) \to \mathscr{F}(\mathscr{H}_{\mathrm{PF}}) \otimes \mathscr{F}(\mathscr{H}_{\mathrm{PF}})$ is also isometry. Here $\mathscr{U}_{\mathrm{c}} : \mathscr{F}(\mathscr{H}_{\mathrm{PF}} \oplus \mathscr{H}_{\mathrm{PF}}) \to \mathscr{F}(\mathscr{H}_{\mathrm{PF}}) \otimes \mathscr{F}(\mathscr{H}_{\mathrm{PF}})$ is the canonical unitary. More specifically it acts as

$$U_P \prod_{j=1}^n a^\dagger(f_j)\Omega = \prod_{j=1}^n (a^\dagger(\hat{j}_{1,P} f_j) \otimes \mathbb{1}_{\mathscr{F}} + \mathbb{1}_{\mathscr{F}} \otimes a^\dagger(\hat{j}_{2,P} f_j))\Omega \otimes \Omega.$$

We give a comment on the localization of a boson Fock space. We set $\mathbb{1}_{L^2} = \mathbb{1}_2$ and $\bar{U}_P = \mathbb{1}_2 \otimes U_P$. For sufficiently smooth F and G, $\mathrm{q}(F, G) = \frac{1}{2}(P_A F, P_A G) + (F, H_{\mathrm{rad}} G) + (F, VG)$. Since $\bar{U}_P$ is isometry, we have

$$\mathrm{q}(F, G) = \frac{1}{2}(\bar{U}_P P_A F, \bar{U}_P P_A G) + (\bar{U}_P F, \bar{U}_P H_{\mathrm{rad}} G) + (F, VG)$$

and the intertwining property of $\bar{U}_P$ we have

$$\begin{aligned} \bar{U}_P P_A &= \left((-i\nabla - A^{(1)}) \otimes \mathbb{1}_{\mathscr{F}} + \mathbb{1}_{\mathscr{H}} \otimes A^{(2)}\right) \bar{U}_P, \\ \bar{U}_P H_{\mathrm{rad}} &= (H_{\mathrm{rad}} \otimes \mathbb{1}_{\mathscr{F}} + \mathbb{1}_{\mathscr{H}} \otimes H_{\mathrm{rad}}) \bar{U}_P - U \mathbb{1}_2 \otimes \mathrm{d}\Gamma(\hat{j}_P, [\omega \oplus \omega, \hat{j}_P]_{\oplus}). \end{aligned}$$

Here $A^{(1)}$ and $A^{(2)}$ are given in (3.8) and (3.9), respectively. Roughly speaking it seems that

$$\left((-i\nabla - A^{(1)})^2 + H_{\mathrm{rad}}\right) \otimes \mathbb{1}_{\mathscr{F}} \sim \left(P_A^2 + H_{\mathrm{rad}}\right) \otimes \mathbb{1}_{\mathscr{F}}$$

and

$$\mathbb{1}_{\mathscr{H}} \otimes ((A^{(2)})^2 + H_{\mathrm{rad}}) \sim \mathbb{1}_{\mathscr{H}} \otimes H_{\mathrm{rad}}$$

as $P \to \infty$. Hence

$$\begin{aligned} \mathrm{q}(\chi_R F, \chi_R G) &\sim (\bar{U}_P \chi_R F, (H_{\mathrm{PF}} \otimes \mathbb{1}_{\mathscr{F}}) \bar{U}_P \chi_R G) + (\bar{U}_P \chi_R F, (\mathbb{1}_{\mathscr{H}} \otimes H_{\mathrm{rad}}) \bar{U}_P \chi_R G) \\ &\sim E_R \|\chi_R F\|^2 + \nu \|\chi_R F\|^2 - \|(\mathbb{1}_{\mathscr{H}} \otimes P_\Omega) \bar{U}_P \chi_R F\|^2. \end{aligned}$$

The key point of this proof is to prove $\|(\mathbb{1}_{\mathscr{H}} \otimes P_\Omega)\bar{U}_P \chi_R F_n\| \to 0$ as $n \to \infty$ for some sequence $\{F_n\}_n^\infty$ by a compactness argument.

Lemma 3.14 *Let $F_n \in Q(\mathfrak{q})$ be such that $\|F_n\| = 1$ and $w - \lim_{n\to\infty} F_n = 0$. Then*

(1) $\mathfrak{q}(\chi_R F_n, \chi_R F_n) \geq (E+\nu)\|\chi_R F_n\|^2 - \nu\|(\mathbb{1}_{\mathscr{H}} \otimes P_\Omega)\bar{U}_P \chi_R F_n\|^2 + o_R(P)$,
(2) $\mathfrak{q}(\tilde{\chi}_R F_n, \tilde{\chi}_R F_n) \geq E_R\|\tilde{\chi}_R F_n\|^2$,
(3) $\lim_{n\to\infty} \|(\mathbb{1}_{\mathscr{H}} \otimes P_\Omega)\bar{U}_P \chi_R F_n\| = 0$.

Here $o_R(P)$ is a function such that $\lim_{P\to\infty} o_R(P) = 0$ for each R. In particular

$$\mathfrak{q}(\chi_R F_n, \chi_R F_n) = (E+\nu)\|\chi_R F_n\|^2 + o(n^0) + o_R(P),$$

where $o(n^0)$ is a function such that $\lim_{n\to\infty} o(n^0) = 0$.

Proof In this proof we set $h = H_{\mathrm{rad}}^{1/2}$, and $h_1 = \mathbb{1}_2 \otimes h \otimes \mathbb{1}_{\mathscr{F}}$ and $h_2 = \mathbb{1}_2 \otimes \mathbb{1}_{\mathscr{F}} \otimes h$. Let $F, G \in C_0^\infty(\mathbb{R}^3) \otimes (\mathscr{F}_{\mathrm{fin}} \cap D(H_{\mathrm{rad}}))$. We have

$$\mathfrak{q}(F, G) = \frac{1}{2}(\bar{U}_P P_A F, \bar{U}_P P_A G) + (\bar{U}_P F, \bar{U}_P H_{\mathrm{rad}} G) + (F, VG).$$

Define

$$\mathfrak{q}_1(F, G) = \frac{1}{2}((P_A \otimes \mathbb{1}_{\mathscr{F}})\bar{U}_P F, (P_A \otimes \mathbb{1}_{\mathscr{F}})\bar{U}_P G) + (h_1\bar{U}_P F, h_1\bar{U}_P G) + (F, VG),$$
$$\mathfrak{q}_2(F, G) = (h_2\bar{U}_P F, h_2\bar{U}_P G).$$

We can show that

$$\mathfrak{q}(\chi_R F_n, \chi_R F_n) = \mathfrak{q}_1(\chi_R F_n, \chi_R F_n) + \mathfrak{q}_2(\chi_R F_n, \chi_R F_n) + o_R(P). \tag{3.7}$$

We shall prove this in Lemma 3.15 below, and this estimate in principle can be derived by estimating commutators $[\bar{U}_P, P_A]$ and $[\bar{U}_P, H_{\mathrm{rad}}]$.

$\mathfrak{q}_1(F, G)$ can be written as $\mathfrak{q}_1(F, G) = (\bar{U}_P F, (H_{\mathrm{PF}} \otimes \mathbb{1}_{\mathscr{F}})\bar{U}_P G)$ for sufficiently smooth F and G and then $\mathfrak{q}_1(F, F) \geq E\|\bar{U}_P F\|^2 = E\|F\|^2$. It is trivial to see that

$$\mathfrak{q}_1(\chi_R F_n, \chi_R F_n) \geq E\|\chi_R F_n\|^2.$$

Since $H_{\mathrm{rad}} \geq \nu(\mathbb{1} - P_\Omega)$, we obtain

$$\mathfrak{q}_2(\chi_R F_n, \chi_R F_n) \geq (\bar{U}_P \chi_R F_n, \nu(\mathbb{1} - P_\Omega)\bar{U}_P \chi_R F_n) = \nu\|\chi_R F_n\|^2 - \nu\|(\mathbb{1}_{\mathscr{H}} \otimes P_\Omega)\bar{U}_P \chi_R F_n\|^2$$

Together with them we have

$$q(\chi_R F_n, \chi_R F_n) = E\|\chi_R F_n\|^2 + \nu\|\chi_R F_n\|^2 - \nu\|(\mathbb{1}_{\mathscr{H}} \otimes P_\Omega)\bar{U}_P \chi_R F_n\|^2 + o_R(P)$$

and (1) follows. (2) is trivial. We prove (3). Since $\bar{U}_P^*(\mathbb{1}_{\mathscr{H}} \otimes P_\Omega)\bar{U}_P = \mathbb{1}_2 \otimes \Gamma(\hat{j}_{1,P})$, it is sufficient to estimate $\|\chi_R \otimes \Gamma(\hat{j}_{1,P})F_n\|$. Let P_N be the projection to the subspace $\oplus_{m=0}^N \mathscr{F}^{(m)}$. We have

$$\|\chi_R \otimes \Gamma(\hat{j}_{1,P})F_n\|^2 = \|\chi_R \otimes \Gamma(\hat{j}_{1,P})P_N F_n\|^2 + \|\chi_R \otimes \Gamma(\hat{j}_{1,P})(\mathbb{1} - P_N)F_n\|^2.$$

We have

$$\|\chi_R \otimes \Gamma(\hat{j}_{1,P})(\mathbb{1} - P_N)F_n\|^2 \le \frac{\|(\mathbb{1}_{\mathscr{H}} \otimes N^{1/2})F_n\|^2}{N}.$$

Notice that

$$\|(\mathbb{1}_{\mathscr{H}} \otimes N^{1/2})F_n\|^2 \le \frac{1}{\nu}\|(\mathbb{1}_{\mathscr{H}} \otimes H_{\rm rad}^{1/2})F\|^2 \le \frac{1}{\nu}q(F_n, F_n).$$

Since $\sup_{n\in\mathbb{N}} q(F_n, F_n) < \infty$, $\sup_{n\in\mathbb{N}} \|(\mathbb{1}_{\mathscr{H}} \otimes N^{1/2})F_n\|^2 < \infty$ is also true. Hence

$$\|\chi_R \otimes \Gamma(\hat{j}_{1,P})(\mathbb{1} - P_N)F_n\|^2 \le \frac{C}{N}$$

with some constant C. We shall estimate $\|\chi_R \otimes \Gamma(\hat{j}_{1,P})P_N F_n\|^2$. Let

$$K = (-\Delta + \mathbb{1})^{-1/4} \otimes (H_{\rm rad} + \mathbb{1})^{-1/4},$$
$$L = \left\{(-\Delta + \mathbb{1})^{1/4} \otimes (H_{\rm rad} + \mathbb{1})^{1/4}\right\}(H_{\rm PF} - E + \mathbb{1})^{-1/2}.$$

We see that $\chi_R \otimes \Gamma(\hat{j}_{1,P})P_N K$ is compact on $L^2(\mathbb{R}_x^3 \times \mathbb{R}_k^{3N})$, and L is bounded. In particular $\chi_R \otimes \Gamma(\hat{j}_{1,P})P_N KL$ is compact. Since

$$\chi_R \otimes \Gamma(\hat{j}_{1,P})P_N F_n = \chi_R \otimes \Gamma(\hat{j}_{1,P})P_N KL(H_{\rm PF} - E + \mathbb{1})^{1/2}F_n$$

and $(H_{\rm PF} - E + \mathbb{1})^{1/2}F_n$ weakly converges to zero as $n \to \infty$, it follows that

$$\lim_{n\to\infty} \|\chi_R \otimes \Gamma(\hat{j}_{1,P})P_N F_n\| = 0.$$

Then (3) follows. ■

Lemma 3.15 *Equation (3.7) is true.*

Proof Firstly we consider $(P_A F, P_A G)$. We have $(P_A F, P_A G) = (\bar{U}_P P_A F, \bar{U}_P P_A G)$ and $\bar{U}_P P_A = ((-i\nabla - A^{(1)}) \otimes \mathbb{1}_{\mathscr{F}} - \mathbb{1}_{\mathscr{H}} \otimes A^{(2)})\bar{U}_P$. Here $A^{(1)}$ and $A^{(2)}$ are localized quantized radiation fields given by

$$A^{(1)}(x) = \frac{1}{\sqrt{2}} \sum_{j=1,2} (a^\dagger(\hat{j}_{1,P} \wp_\mu(x, j), j) + a(\hat{j}_{1,P} \tilde{\wp}_\mu(x, j), j)), \tag{3.8}$$

$$A^{(2)}(x) = \frac{1}{\sqrt{2}} \sum_{j=1,2} (a^\dagger(\hat{j}_{2,P} \wp_\mu(x, j), j) + a(\hat{j}_{2,P} \tilde{\wp}_\mu(x, j), j)). \tag{3.9}$$

We estimate

$$\begin{aligned} Q &= (-i\nabla - A) \otimes \mathbb{1}_{\mathscr{F}} - \left\{(-i\nabla - A^{(1)}) \otimes \mathbb{1}_{\mathscr{F}} - \mathbb{1}_{\mathscr{H}} \otimes A^{(2)}\right\} \\ &= -\left(\sum_{j=1,2} (a^\dagger((\mathbb{1} - \hat{j}_{1,P}) \wp_\mu(x, j), j) + a((\mathbb{1} - \hat{j}_{1,P}) \tilde{\wp}_\mu(x, j), j)) \right) \otimes \mathbb{1}_{\mathscr{F}} \\ &\quad + \mathbb{1}_{\mathscr{H}} \otimes \left(\sum_{j=1,2} (a^\dagger(\hat{j}_{2,P} \wp_\mu(x, j), j) + a(\hat{j}_{2,P} \tilde{\wp}_\mu(x, j), j)) \right). \end{aligned}$$

We have

$$\|a^\sharp((\mathbb{1} - \hat{j}_{1,P}) \wp_\mu(x, j), j) \bar{U}_P \chi_R F\| \le \|(\mathbb{1} - \hat{j}_{1,P}) \wp_\mu(x, j) \chi_R\| \|(N + \mathbb{1})^{1/2} \bar{U}_P F\|$$

and

$$\begin{aligned} &\lim_{P\to\infty} \sup_{|x|\le R} \|(\mathbb{1} - \hat{j}_{1,P}) \wp_\mu(x, j) \chi_R\|^2 \\ &= \lim_{P\to\infty} \sup_{|x|\le R} \chi_R(x) \int_{\mathbb{R}^3} |1 - j_{1,P}(y)|^2 |\check{\wp}_\mu(y - x, j)|^2 dy = 0. \end{aligned}$$

Then $\|a^\dagger((\mathbb{1} - \hat{j}_{1,P}) \wp_\mu(x, j), j) \bar{U}_P \chi_R F\| \to 0$ as $P \to \infty$. Similarly we also see that $\lim_{P\to\infty} \|(a^\dagger(\hat{j}_{2,P} \wp_\mu(x, j), j) + a(\hat{j}_{2,P} \tilde{\wp}_\mu(x, j), j)) \chi_R F\| = 0$.

Next we consider $(F, H_{\rm rad} F) = (\bar{U}_P F, \bar{U}_P H_{\rm rad} F)$. We have

$$\bar{U}_P H_{\rm rad} = (H_{\rm rad} \otimes \mathbb{1} + \mathbb{1} \otimes H_{\rm rad}) \bar{U}_P - U_P d\Gamma(\hat{j}_P, [\omega \oplus \omega, \hat{j}_P]_\oplus).$$

We shall estimate $(F, U_P d\Gamma(\hat{j}_P, [\omega \oplus \omega, \hat{j}_P]_\oplus) F)$. Since

$$[\omega \oplus \omega, \hat{j}_P]_\oplus f = [\omega, \hat{j}_{1,P}] f \oplus [\omega, \hat{j}_{2,P}] f,$$

we consider $[\omega, \hat{j}_{1,P}]$ and $[\omega, \hat{j}_{2,P}]$ separately. Note that

$$\omega \in C^1(\mathbb{R}^3), \quad \nabla_\mu \omega \in L^\infty(\mathbb{R}^3), \quad \mu = 1, 2, 3, \tag{3.10}$$

because of $\nu > 0$, and we have $\hat{j}_{1,P} f(k) = (2\pi)^{-3/2} \int_{\mathbb{R}^3} \check{j}_{1,P}(k - x) f(x) dx$, where $\check{j}_{1,P}$ denotes the inverse Fourier transform of $j_{1,P}$. Since $\check{j}_{1,P}(x) = \check{j}_1(Px) P^3$, we have

$$\omega\hat{j}_{1,P}f(k) = (2\pi)^{-3/2}\int_{\mathbb{R}^3}\check{j}_1(x)\omega(k)f(k+\frac{x}{P})dx,$$
$$\hat{j}_{1,P}\omega f(k) = (2\pi)^{-3/2}\int_{\mathbb{R}^3}\check{j}_1(x)\omega(k+\frac{x}{P})f(k+\frac{x}{P})dx.$$

It is straightforward to see that

$$|\omega\hat{j}_{1,P}f(k) - \hat{j}_{1,P}\omega f(k)| \leq (2\pi)^{-3/2}M\int_{\mathbb{R}^3}|\check{j}_1(x)||\frac{x}{P}||f(k+\frac{x}{P})|dx.$$

Here we used that

$$|\omega(k+\frac{x}{P}) - \omega(k)| \leq \sup_{x\in\mathbb{R}^3}|\nabla\omega(x)||\frac{x}{P}|$$

and $M = \sup_{x\in\mathbb{R}^3}|\nabla\omega(x)|$. Hence

$$|\omega\hat{j}_{1,P}f(k) - \hat{j}_{1,P}\omega f(k)| \leq \frac{M(2\pi)^{-3/2}}{P}\|\langle x\rangle^3\check{j}_1\|\|\langle x\rangle^{-2}f(k+\frac{x}{P})\|$$

and we have

$$\int_{\mathbb{R}^3}|\omega\hat{j}_{1,P}f(k) - \hat{j}_{1,P}\omega f(k)|^2dk \leq \frac{M^2(2\pi)^{-3}}{P^2}\int_{\mathbb{R}^3}\|\langle x\rangle^3\check{j}_1\|^2\|\langle x\rangle^{-2}f(k+\frac{x}{P})\|^2dk \leq \frac{C^2}{P^2}\|f\|^2,$$

where $C = (2\pi)^{-3/2}M\|\langle x\rangle^3\check{j}_1\|\|\langle x\rangle^{-2}\|$. It follows that $\|[\omega, \hat{j}_{1,P}]f\| \leq C\|f\|/P$. Since $[\omega, \hat{j}_{2,P}] = -[\omega, \mathbb{1} - \hat{j}_{2,P}]$, we can see that $\|[\omega, \mathbb{1} - \hat{j}_{2,P}]f\| \leq C\|f\|/P$ in the same way as $\|[\omega, \hat{j}_{1,P}]f\| \leq C\|f\|/P$. By Proposition 2.26 we have

$$(F, d\Gamma(\hat{j}_P, [\omega\oplus\omega, \hat{j}_P]_\oplus)F) \leq \frac{C}{P}\|N^{1/2}F\|^2$$

and the lemma follows. ■

Theorem 3.16 (Griesemer, Lieb and Loss [21, Theorem 4.1]) *Suppose $\nu > 0$ and Assumption 3.10. Then H_{PF} has a ground state.*

Proof Let $F_n \in Q(\mathrm{q})$ be such that $\|F_n\| = 1$ and $w - \lim_{n\to\infty} F_n = 0$. It is sufficient to show that $\liminf_{n\to\infty} q(F_n, F_n) > E$ by Proposition 3.9. We have the identity

$$\mathrm{q}(F, F) = \mathrm{q}(\tilde{\chi}_R F, \tilde{\chi}_R F) + \mathrm{q}(\chi_R F, \chi_R F) - \frac{1}{2}\||\nabla\chi_R|F\|^2 - \frac{1}{2}\||\nabla\tilde{\chi}_R|F\|^2.$$

Since $\||\nabla\chi_R|F\|^2 + \||\nabla\tilde{\chi}_R|F\|^2 \leq C/R^2$, we have by Lemma 3.14

$$\begin{aligned}\liminf_{n\to\infty} \mathrm{q}(F_n, F_n) &\geq \liminf_{n\to\infty} \left((E+\nu)\|\chi_R F_n\|^2 + E_R\|\tilde{\chi}_R F_n\|^2\right) + o_R(P) \\ &= E + \liminf_{n\to\infty} \left(\nu\|\chi_R F_n\|^2 + (E_R - E)\|\tilde{\chi}_R F_n\|^2\right) + o_R(P) \\ &\geq E + \min\{\nu, E_R - E\} + o_R(P).\end{aligned}$$

Taking $P \to \infty$ on both sides above, we have

$$\liminf_{n\to\infty} \mathrm{q}(F_n, F_n) \geq E + \min\{\nu, E_R - E\}.$$

Taking $R \to \infty$ again we have

$$\liminf_{n\to\infty} \mathrm{q}(F_n, F_n) \geq E + \min\{\nu, \Sigma - E\}.$$

Since $\min\{\nu, \Sigma - E\} > 0$, $\liminf_{n\to\infty} \mathrm{q}(F_n, F_n) > E$ and the lemma is proven. ■

3.4 Existence of the Ground State for Massless Case $\nu = 0$

In this section we show the existence of the ground state of H_{PF} with $\nu = 0$. To see this we need several localization estimates on position x of the quantum matter, the number of photons and momentum k of a photon.

3.4.1 *Spatial Localization*

Under the binding condition it can be shown that $\|e^{\beta|x|}E_{(-\infty,\lambda]}\| < \infty$ for some sufficiently small $\beta > 0$, where E_O is the spectral projection of H_{PF} for $O \subset \mathbb{R}$. In particular $\|e^{\beta|x|}\Psi_{\mathrm{g}}\|_{\mathscr{H}} < \infty$ follows. This is shown by Bach, Fröhlich and Sigal [8, Theorem II.1] and Griesemer [20, Theorem 1]. In [20] a sufficient condition of spacial exponential decays for general Hamiltonians is given in an abstract setting.

Proposition 3.17 ([7, Theorem I. II], [20, Theorem 1]) *Suppose Assumption 3.10. Let E be the ground state energy of H_{PF} and $\Sigma = \lim_{R\to\infty} E_R$ the ionization energy. Assume that $\lambda \geq E$ and $\beta > 0$ such that $\lambda + \beta^2 < \Sigma$. Then $\|e^{\beta|x|}E_{(-\infty,\lambda]}\| < \infty$.*

Proof This proof is due to Griesemer [20, Theorem 1]. Let $(j_1, j_2) \in l_{\mathrm{loc}}(\mathbb{R}^3)$. Define $H_R = H_{\mathrm{PF}} + (E_R - E)\chi_{2R}$. By IMS localization formula we have

$$(F, H_R F) = \sum_{i=1}^{2} (F, j_i H_R j_i F) - \|\nabla j_1\|^2\|F\| - \|\nabla j_2\|^2\|F\| \geq E_R - \frac{C}{R^2}\|F\|^2.$$

It follows that

$$H_R \geq E_R - \frac{C}{R^2}.$$

Let R be sufficiently large such that $\lambda + \beta^2 < E_R - C/R^2$ and we set

$$\delta = E_R - \frac{C}{R^2} - \beta^2 - \lambda > 0.$$

Let $g \in C_0^\infty(\mathbb{R})$ be such that

$$g(x) = \begin{cases} 1, & x \in [E, \lambda], \\ 0, & x \in [\lambda + \delta/2, \infty), \end{cases}$$

and $f(x) = \frac{\beta\langle x\rangle}{1+\varepsilon\langle x\rangle}$ with $\varepsilon > 0$. $\|\nabla f\|_\infty \leq \beta < \infty$ uniformly in $\varepsilon > 0$. We shall show that $e^f g(H)$ is bounded. Let $\tilde{g}$ be an almost analytic extension[1] of g, and

$$\begin{aligned} g(H) &= -\frac{1}{\pi}\int_{\mathbb{C}} \frac{\partial \tilde{g}}{\partial \bar{z}}(z-H)^{-1}dxdy = -\frac{1}{\pi}\int_{\mathbb{C}} \frac{\partial \tilde{g}}{\partial \bar{z}}((z-H)^{-1} - (z-H_R)^{-1})dxdy \\ &= -\frac{1}{\pi}\int_{\mathbb{C}} \frac{\partial \tilde{g}}{\partial \bar{z}}((z-H_R)^{-1}(E_R - E)\chi_{2R}(z-H_R)^{-1})dxdy. \end{aligned}$$

Here we used $g(H_R) = 0$, since $\operatorname{supp} g \subset [E, \lambda + \delta/2)$, $\operatorname{sp}(H_R) \subset [E_R - C^2/R, \infty)$ and $[E, \lambda + \delta/2) \cap [E_R - C^2/R, \infty) = \emptyset$. Thus

$$\|e^f g(H)\| = \sup_{z \in \operatorname{supp}\tilde{g}} \|e^f (z - H_R)^{-1} e^{-f}\| \|e^f \chi_{2R}\|_\infty \frac{(E_R - E)}{\pi} \int_{\mathbb{C}} \left|\frac{\partial \tilde{g}}{\partial \bar{z}}\right| \|(z-H)^{-1}\| dxdy.$$

Since $e^f(z - H_R)^{-1}e^{-f} = (z - e^f H_R e^{-f})^{-1}$ and

$$\begin{aligned} 2\Re(F, e^f H_R e^{-f} F) &= (e^{-f}F, (e^{2f}H_R + H_R e^{2f})e^{-f}F) \\ &= 2(e^{-f}F, e^f H_R e^f e^{-f} F) - 2\|\nabla f e^f e^{-f} F\|^2, \end{aligned}$$

we have $2\Re(F, e^f H_R e^{-f} F) = 2(F, (H_R - |\nabla f|^2)F)$ and it follows that

$$\Re(F, (e^f H_R e^{-f} - z)F) \geq E_R - \frac{C}{R^2} - \beta^2 - \Re z \geq \frac{\delta}{2}$$

for $z \in \operatorname{supp}\tilde{g}$. Hence $\|(e^f H_R e^{-f} - z)F\| \geq \delta\|F\|/2$ which implies that

$$\|(z - e^f H_R e^{-f})^{-1}\| < \frac{2}{\delta}$$

---

[1]Suppose $f(x) \in C_0^\infty(\mathbb{R}^d)$. Then there exists a function $F(z) \in C_0^\infty(\mathbb{C})$ such that (1) $F(x) = f(x)$ for $x \in \mathbb{R}^d$, (2) $|\partial_{\bar{z}} F(z)| \leq C_n |\Im z|^n$ for any $n \in \mathbb{N}$, (3) $f(A) = \frac{1}{\pi}\int_{\mathbb{C}} \partial_{\bar{z}} F(z)(A-z)^{-1}dxdy$, where $\partial_{\bar{z}} = \partial_x + i\partial_y$. The integral is defined in the uniform operator topology. F is called almost analytic extension of f. We refer to see [26].

for $z \in \mathrm{supp}\tilde{g}$. It is proven that

$$\|e^f g(H)\| \le \frac{2}{\delta}\|e^f \chi_{2R}\|_\infty (E_R - E)\frac{1}{\pi}\int_{\mathbb{C}} \left|\frac{\partial \tilde{g}}{\partial \bar{z}}\right| \|(z-H)^{-1}\| dxdy$$

and in particular $\|e^f E_{(-\infty,\lambda]}\| < \infty$. Take $\varepsilon \downarrow 0$. Then the lemma is proven. ∎

3.4.2 *Pointwise Exponential Decay*

We can show the pointwise exponential decay of bound state of H_{PF} when we assume suitable conditions on V. Let $f, g \in L^2(\mathbb{R}^3)$. We have

$$(f, e^{-tH_{\mathrm{p}}} g) = \int_{\mathbb{R}^3} dx \mathbb{E}^x \left[\bar{f}(x) g(B_t) e^{-\int_0^t V(B_s)ds} \right].$$

Here $(B_t)_{t\ge 0}$ is 3-dimensional Brownian motion starting in x at $t = 0$ on a probability space $(S, \mathscr{F}, P^x)$, and $\mathbb{E}^x$ denotes the expectation with respect to P^x. This is the so-called Feynman-Kac formula. Let us consider the Schrödinger equation of the form

$$H_{\mathrm{p}}\Psi_{\mathrm{p}} = E_{\mathrm{p}}\Psi_{\mathrm{p}}.$$

By applying Feynman-Kac formula to identity $\Psi_{\mathrm{p}} = e^{tE_{\mathrm{p}}} e^{-tH_{\mathrm{p}}}\Psi_{\mathrm{p}}$, we can prove pointwise spatially exponential decay of bound state Ψ_{p}. This method can be also applicable to show the spatially exponential decay of bound states of the Pauli–Fierz Hamiltonian. We shall show only the outline. We introduce classes of external potentials.

Definition 3.18 (*External potentials*) We say that $V \in \mathscr{V}$ if and only if $V = W - U$ such that

(1) $U \ge 0$ and $U \in L^p(\mathbb{R}^3)$ for $p > d/2$ and $1 \le p < \infty$,
(2) $W \in L^1_{\mathrm{loc}}(\mathbb{R}^3)$ and $W_\infty = \inf_{x\in\mathbb{R}^3} W(x) > -\infty$.

For $V = W - U \in \mathscr{V}$ we define $V_\infty = \liminf_{|x|\to\infty} V(x)$. Since $U \in L^p(\mathbb{R}^3)$, we see that $\liminf_{|x|\to\infty} U(x) = 0$ and hence $V_\infty = \liminf_{|x|\to\infty} W(x)$. Moreover $V_\infty \ge W_\infty$ holds.

Let Φ be a bound state of H_{PF} and $H_{\mathrm{PF}}\Phi = E_{\mathrm{b}}\Phi$. Since $\Phi = e^{tE_{\mathrm{b}}} e^{-tH_{\mathrm{PF}}}\Phi$, the functional integral representation yields that

$$\|\Phi(x)\|_{\mathscr{F}} \le \mathbb{E}^x[e^{-\int_0^t V(B_s)ds} \|\Phi(B_t)\|_{\mathscr{F}}]. \tag{3.11}$$

See e.g. [31] for the derivation of (3.11).

Lemma 3.19 (Carmona estimate) *Suppose that $V \in \mathscr{V}$. Then for any $t, a > 0$ and every $0 < \alpha < 1/2$, there exist constants $D_1, D_2, D_3 > 0$ such that*

$$\|\Phi(x)\|_{\mathscr{F}} \le t^{-3/2} D_1 e^{D_2 \|U\|_p^m t} e^{E_b t} (D_3 e^{-\frac{\alpha}{4}\frac{a^2}{t}} e^{-tW_\infty} + e^{-tW_a(x)}) \|\Phi\|_{\mathscr{H}}, \quad (3.12)$$

where $W_a(x) = \inf\{W(y) | |x - y| < a\}$ and $m = (1 - 3/(2p))^{-1}$.

Proof The proof is a modification of [12]. By Schwarz inequality, we have

$$\|\Phi(x)\|_{\mathscr{F}} \le e^{tE_b} (\mathbb{E}^x[e^{-4\int_0^t W(B_s)ds}])^{1/4} (\mathbb{E}^x[e^{+4\int_0^t U(B_s)ds}])^{1/4} \mathbb{E}^0[\|\Phi(x + B_t)\|^2]^{1/2}. \quad (3.13)$$

Note that $\mathbb{E}^0[\|\Phi(x + B_t)\|^2] \le t^{-3/2} \|\Phi\|^2$. Let $A = \{w \in S | \sup_{0 \le s \le t} |B_s(w)| > a\}$. It follows from Lévy's maximal inequality that

$$\mathbb{E}^0[1_A] \le 2\mathbb{E}^0[1_{\{|B_t| \ge a\}}] = 2(2\pi)^{-3/2} 4\pi \int_{a/\sqrt{t}}^\infty e^{-r^2/2} r^2 dx \le \xi_\alpha e^{-\alpha a^2/t}$$

with some ξ_α, for every $0 < \alpha < 1/2$. The first factor in (3.13) is estimated as

$$\mathbb{E}^x[e^{-4\int_0^t W(B_s)ds}] = \mathbb{E}^0[1_A e^{-4\int_0^t W(B_s + x)ds}] + \mathbb{E}^x[1_{A^c} e^{-4\int_0^t W(B_s)ds}]$$
$$\le e^{-4tW_\infty} \mathbb{E}^0[1_A] + e^{-4tW_a(x)} \le \xi_\alpha e^{-\alpha a^2/t} e^{-4tW_\infty} + e^{-4tW_a(x)}.$$

Next we estimate the second factor. Since U is in Kato-class, there exist constants $D_1, D_2 > 0$ such that

$$\mathbb{E}^x[e^{-4\int_0^t U(B_s)ds}] \le D_1 e^{D_2 \|U\|_p^m t}.$$

Setting $D_3 = \xi_\alpha^{1/4}$, we obtain the lemma by using $(a + b)^{1/4} \le a^{1/4} + b^{1/4}$ for $a, b \ge 0$. ■

Proposition 3.20 *Let $V = W - U \in \mathscr{V}$.*

(1) Suppose that $W(x) \ge \gamma |x|^{2n}$ outside a compact set K, for some $n > 0$ and $\gamma > 0$. Take $0 < \alpha < 1/2$. Then there exists a constant $C_1 > 0$ such that

$$\|\Phi(x)\|_{\mathscr{F}} \le C_1 \exp\left(-\frac{\alpha c}{16} |x|^{n+1}\right) \|\Phi\|_{\mathscr{H}},$$

where $c = \inf_{x \in \mathbb{R}^3 \setminus K} W_{|x|/2}(x)/|x|^{2n}$.

(2) Suppose that $V_\infty > E_b$, $V_\infty > W_\infty$, and $0 < \beta < 1$. Then there exists a constant $C_2 > 0$ such that

$$\|\Phi(x)\|_{\mathscr{F}} \le C_2 \exp\left(-\frac{\beta}{8\sqrt{2}} \frac{(V_\infty - E_b)}{\sqrt{V_\infty - W_\infty}} |x|\right) \|\Phi\|_{\mathscr{H}}.$$

(3) *Suppose that* $\lim_{|x|\to\infty} W(x) = \infty$. *Then there exist constants* $C, \delta > 0$ *such that*

$$\|\Phi(x)\|_{\mathscr{F}} \leq C \exp(-\delta|x|)\, \|\Phi\|_{\mathscr{H}}.$$

Proof Since $\sup_{x\in\mathbb{R}^3} \|\Phi(x)\|_{\mathscr{F}} < \infty$, it suffices to show all the statements for sufficiently large $|x|$. Note that $W_{|x|/2}(x) \geq c|x|^{2n}$ for $x \in \mathbb{R}^3 \setminus K$. We have the following bounds for $x \in \mathbb{R}^3 \setminus K$:

$$|x| W_{|x|/2}(x)^{1/2} \geq c|x|^{n+1}, \quad |x| W_{|x|/2}(x)^{-1/2} \leq c|x|^{1-n}.$$

Inserting $t = t(x) = |x| W_{|x|/2}(x)^{-1/2}$ and $a = a(x) = |x|/2$ in (3.12), we have

$$\|\Phi(x)\|_{\mathscr{F}} \leq e^{-\frac{\alpha}{16}c|x|^{n+1}} D_1 e^{(D_2\|U\|_p^m + E_b)c|x|^{1-n}} (D_3 e^{c|x|^{1-n}|W_\infty|} + e^{-(1-\frac{\alpha}{16})c|x|^{n+1}})\|\Phi\|$$

for $x \in \mathbb{R}^3 \setminus K$, where $m = (1 - \frac{d}{2p})^{-1}$. Hence (1) follows. Rewrite formula (3.12) as

$$\|\Phi(x)\|_{\mathscr{F}} \leq D_1 e^{D_2\|U\|_p^m t} (D_3 e^{-\frac{\alpha}{4}\frac{a^2}{t}} e^{-t(W_\infty - E_b)} + e^{-t(W_a(x) - E_b)})\|\Phi\|. \tag{3.14}$$

Then on flipping signs, with $V_\infty = \liminf_{|x|\to\infty}(-W_-(x))$ and $V_\infty > W_\infty$ it is possible to choose a decomposition $V = W - U \in \mathscr{V}$ such that $D_2\|U\|_p^m \leq (V_\infty - E_b)/2$ since $\liminf_{|x|\to\infty}(-U(x)) = 0$. Inserting $t = t(x) = \varepsilon|x|$ and $a = a(x) = |x|/2$ in (3.14), we have

$$\|\Phi(x)\|_{\mathscr{F}} \leq D_1'(e^{-(\frac{\alpha}{16\varepsilon} + \varepsilon(W_\infty - E_b) - \frac{1}{2}\varepsilon(V_\infty - E_b))|x|} + e^{-\varepsilon((W_{|x|/2}(x) - E_b) - \frac{1}{2}(V_\infty - E_b))|x|})\|\Phi\|.$$

Choosing $\varepsilon = \sqrt{\alpha/16}/\sqrt{V_\infty - W_\infty}$, the exponent in the first term above becomes

$$\frac{\alpha}{16\varepsilon} + \varepsilon(W_\infty - E_b) - \frac{1}{2}\varepsilon(V_\infty - E_b) = \frac{1}{2}\varepsilon(V_\infty - E_b).$$

Moreover, we see that $\liminf_{|x|\to\infty} W_{|x|/2}(x) = V_\infty$, and obtain

$$\|\Phi(x)\|_{\mathscr{F}} \leq C_2 e^{-\frac{\varepsilon}{2}(V_\infty - E_b)|x|}\|\Phi\|$$

for sufficiently large $|x|$. Thus (2) follows. In the confining case, for any $c > 0$ there exists $N > 0$ such that $W_{|x|/2}(x) \geq c$, for all $|x| > N$. Inserting $t = t(x) = \varepsilon|x|$ and $a = a(x) = |x|/2$ in (3.12), we obtain that

$$\|\Phi(x)\|_{\mathscr{F}} \leq D_1'(e^{-(\frac{\alpha}{16\varepsilon} - \varepsilon D_2\|U\|_p^m + \varepsilon(W_\infty - E_b))|x|} + e^{-\varepsilon(c - E_b - D_2\|U\|_p^m)|x|})\|\Phi\|$$

for $|x| > N$. Choosing sufficiently large c and sufficiently small ε such that

$$\frac{\alpha}{16\varepsilon} - \varepsilon D_2 \|U\|_p^m + \varepsilon(W_\infty - E_b) > 0 \quad \text{and} \quad c - E_b - D_2\|U\|_p^m > 0,$$

we obtain $\|\Phi(x)\|_{\mathscr{F}} \leq C' e^{-\delta'|x|}$ for large enough $|x|$. Thus (3) follows. ■

3.4.3 Carleman Operator and Pull-Through Formula

Let Ψ_ν be the ground state of $H_{\rm PF}$ with positive mass $\nu > 0$. In this section we shall see the boson number localization of massive ground state Ψ_ν.

To show this we need the so-called pull-through formula which can be derived in several ways. The pull-through formula is a tool to give an estimate of $\|N^{\frac{1}{2}}\Psi_\nu\|$. Formal but the most used prescription to derive the pull-through formula is as follows. Using formal commutation relations $[a(k,j), a^\dagger(k',j')] = \delta_{kk'}\delta_{jj'}$, $[a(k,j), H_{\rm rad}] = -\omega(k)a(k,j)$, and $H_{\rm PF}\Psi_\nu = E\Psi_\nu$, we have

$$(H_{\rm PF} - E + \omega(k))a(k,j)\Psi_\nu = (-i\nabla - A)\frac{e(k,j)\hat{\varphi}(k)}{\sqrt{2\omega(k)}}e^{-ikx}\Psi_\nu$$

and

$$a(k,j)\Psi_\nu = (H_{\rm PF} - E + \omega(k))^{-1}(-i\nabla - A)\frac{e(k,j)\hat{\varphi}(k)}{\sqrt{2\omega(k)}}e^{-ikx}\Psi_\nu.$$

Since $\|N^{1/2}\Psi_\nu\|^2 = \sum_{j=1,2}\int_{\mathbb{R}^3}\|a(k,j)\Psi_\nu\|^2 dk$, we can see that

$$\|N^{1/2}\Psi_\nu\|^2 = \sum_{j=1,2}\int_{\mathbb{R}^3}\left\|(H_{\rm PF} - E + \omega(k))^{-1}(-i\nabla - A)\frac{e(k,j)\hat{\varphi}(k)}{\sqrt{2\omega(k)}}e^{-ikx}\Psi_\nu\right\|^2 dk.$$

In this section however we shall derive pull-through formula by a combination of asymptotic fields and a Carleman operator. Define

$$a_t(f,j) = e^{-itH_{\rm PF}}e^{itH_{\rm rad}}a(f,j)e^{-itH_{\rm rad}}e^{itH_{\rm PF}}, \quad t \in \mathbb{R}.$$

Since $a_t(f,j)\Psi_\nu = e^{-it(H_{\rm PF}-E)}a(e^{-it\omega}f,j)\Psi_\nu$, we can see that

$$\text{s-}\lim_{t\to\pm\infty} a_t(f,j)\Psi_\nu = 0$$

by the Riemann–Lebesgue lemma. By this and the identity

$$(F, a_t(f,j)\Psi_\nu) - (F, a(f,j)\Psi_\nu) = \int_0^t \frac{d}{ds}(F, a_s(f,j)\Psi_\nu)ds,$$

we can also see that

$$0 = \lim_{t\to+\infty} (F, a_t(f, j)\Psi_\nu)$$
$$= (F, a(f, j)\Psi_\nu) + i \int_{\mathbb{R}^3} dk \int_0^\infty (F, f(k)e^{-is(H_{\mathrm{PF}}-E+\omega(k))} D_j\Psi_\nu)ds.$$

Here

$$D_j = (-i\nabla - A)e(k, j)e^{-ikx} \frac{\hat{\varphi}(k)}{\sqrt{2\omega(k)}}.$$

Let $f, f/\sqrt{\omega} \in L^2(\mathbb{R}^3)$. We have [35, Lemma 2.7]

$$a(f, j)\Psi_\nu = -\int_{\mathbb{R}^3} f(k)\kappa_j(k)dk,$$

where the integral kernel is given by

$$\kappa_j(k) = (H_{\mathrm{PF}} - E + \omega(k))^{-1} D_j\Psi_\nu.$$

Let $T_{g,j} : L^2(\mathbb{R}^3) \to \mathscr{H}$ be defined by

$$T_{g,j} f = -\int_{\mathbb{R}^3} f(k)\kappa_j(k)dk.$$

$T_{g,j}$ is an integral operator with $\mathscr{H}$-valued kernel, and $a(f, j)\Psi_\nu = -T_{g,j} f$ follows. Adjoint

$$T^*_{g,j} : \mathscr{H} \ni F \mapsto -(\kappa_j(\cdot), F)_{\mathscr{H}} \in L^2(\mathbb{R}^3)$$

is a Carleman operator with the domain $D(T^*_{g,j}) = \{F \in \mathscr{F} | (\kappa_j(\cdot), F)_{\mathscr{H}} \in L^2(\mathbb{R}^3)\}$. Thus $T^*_{g,j}$ is Hilbert–Schmidt if and only if $\|\kappa_j(\cdot)\|_{\mathscr{H}}$ is an L^2-function, and

$$\|T^*_{g,j}\|_2^2 = \int_{\mathbb{R}^3} \|\kappa_j(k)\|^2_{\mathscr{H}} dk < \infty$$

follows. Here $\|\cdot\|_2$ denotes the Hilbert–Schmidt norm. Hence $\int_{\mathbb{R}^3} \|\kappa_j(k)\|^2 dk < \infty$ if and only if $T_{g,j}$ is a Hilbert–Schmidt operator. Furthermore by the definition of the Hilbert–Schmidt norm we see that $\|T_{g,j}\|_2^2 = \sum_{m=1}^\infty \|T_{g,j}e_m\|^2_{\mathscr{H}}$ for any complete orthonormal basis $\{e_m\}_{m=1}^\infty$ in $L^2(\mathbb{R}^3)$. It is noticed that

$$\sum_{j=1,2}\sum_{m=1}^\infty \|T_{g,j}e_m\|^2_{\mathscr{H}} = \sum_{j=1,2}\sum_{m=1}^\infty \|a(e_m, j)\Psi_\nu\|^2_{\mathscr{H}} = \|N^{1/2}\Psi_\nu\|^2_{\mathscr{H}}.$$

Here the second identity is straightforwardly derived.

Proposition 3.21 ([35, Theorem 2.9]) $\Psi_\nu \in D(N^{1/2})$ *if and only if*

$$\sum_{j=1,2} \int_{\mathbb{R}^3} \|\kappa_j(k)\|^2_{\mathscr{H}} dk < \infty$$

and when $\Psi_\nu \in D(N^{1/2})$, it follows that

$$\|N^{1/2}\Psi_\nu\|^2 = \sum_{j=1,2} \int_{\mathbb{R}^3} \|\kappa_j(k)\|^2_{\mathscr{H}} dk. \tag{3.15}$$

Equation (3.15) is called the pull-through formula.

3.4.4 Regularized Pull-Through Formula

By the pull-through formula derived in Sect. 3.4.3 we immediately see that

$$\|N^{1/2}\Psi_\nu\| \leq \int_{\mathbb{R}^3} \frac{|\hat{\varphi}(k)|^2}{2\omega(k)^3} \sum_{j=1,2} \|(-i\nabla - A)e(k,j)\Psi_\nu\|^2 dk$$

by the Coulomb gauge condition. Let $T_\mu = -i\nabla_\mu - A_\mu$. Since we have the bound

$$\begin{aligned}\sum_{j=1,2} \|(-i\nabla - A)e(k,j)\Phi\|^2 &= (\Phi, T^2\Phi) - \sum_{\mu \neq \nu}^{3} \frac{k_\mu k_\nu}{|k|^2}(T_\mu\Phi, T_\nu\Phi) \\ &\leq C_1(\Phi, T^2\Phi) \leq C_2(\Phi, H_{\mathrm{PF}}\Phi) + C_3\|\Phi\|^2,\end{aligned}$$

we have

$$\sum_{j=1,2} \|(-i\nabla - A)e(k,j)\Phi\|^2 \leq (C'E + C'')\|\Psi_\nu\|^2,$$

$$\|N^{1/2}\Psi_\nu\|^2 \leq C_4 \int_{\mathbb{R}^3} \frac{|\hat{\varphi}(k)|^2}{\omega(k)^3} dk \tag{3.16}$$

for $\|\Psi_\nu\| = 1$. As $\nu \to 0$, the right-hand side above diverges if $\hat{\varphi}(0) \neq 0$, and (3.16) is not sufficient to show the bound on $\|N^{1/2}\Psi_\nu\|$ uniformly in $\nu > 0$.

Hence we introduce a regularized Hamiltonian to get the boson number localization which is uniform in ν. The crucial idea is to use that $\||x|^n\Psi_\nu\|$ has an upper bound uniformly in ν, which was shown in the previous section, and $|e^{-ikx} - 1| \leq |k||x|$.

The Pauli–Fierz Hamiltonian H_{PF} can be transformed as

$$H_{\mathrm{PF}}^{\mathrm{reg}} = e^{-ixA(0)} H_{\mathrm{PF}} e^{ixA(0)} = (-i\nabla - A^R(x))^2 + H_{\mathrm{rad}} + h + V,$$

where

$$A_\mu^R(x) = \sum_{j=1,2} \frac{1}{\sqrt{2}} \int e_\mu(k,j) \frac{\hat{\varphi}(k)}{\sqrt{\omega(k)}} \left((e^{-ikx} - 1)a^\dagger(k,j) + (e^{ikx} - 1)a(k,j)\right) dk$$

and $h = \int_{\mathbb{R}^3}^{\oplus} h(x)dx$ with

$$h(x) = \frac{i}{\sqrt{2}} \sum_{j=1,2} \int x \cdot e(k,j)\sqrt{\omega(k)}\hat{\varphi}(k) \left(a^\dagger(k,j) - a(k,j)\right) dk - \frac{1}{2}\|x \cdot e(\cdot,j)\hat{\varphi}\|^2.$$

Formally this transformation is the shift and we have

$$e^{-ixA(0)} H_{\rm rad} e^{ixA(0)} = \sum_{j=1,2} \int \omega(k) b_j^\dagger(k,x) b_j(k,x) dk,$$

where $b_j(k,x) = a(k,j) - i\frac{\hat{\varphi}}{\sqrt{\omega}} e(k,j)x$ for each $x \in \mathbb{R}^3$. From now on we treat $H_{\rm PF}^{\rm reg}$ instead of $H_{\rm PF}$, and both are unitary equivalent which implies that $H_{\rm PF}$ admits the ground state if and only if $H_{\rm PF}^{\rm reg}$ does. So we shall prove the existence of the ground state for $H_{\rm PF}^{\rm reg}$ instead of for $H_{\rm PF}$. Let Ψ_ν be the ground state of $H_{\rm PF}^{\rm reg}$, where we use the same notation as that of $H_{\rm PF}$. In the same procedure as in Sect. 3.4.3 for $H_{\rm PF}$ replaced with $H_{\rm PF}^{\rm reg}$ we can see that

$$0 = (F, a(f,j)\Psi_\nu) + i \int_{\mathbb{R}^3} dk \int_0^\infty ds (F, f(k) e^{-is(H_{\rm PF}^{\rm reg} - E + \omega(k))} C_j(k)\langle x \rangle \Psi_\nu).$$

Here $\langle x \rangle = \sqrt{|x|^2 + 1}$ and

$$C_j(k) = \frac{\hat{\varphi}(k)}{\sqrt{2\omega(k)}} e(k,j) \cdot (-i\nabla - A^R) \frac{(e^{-ikx} - 1)}{\langle x \rangle} + \frac{\rho_j(k)}{\langle x \rangle},$$

$$\rho_j(k) = -i \frac{\sqrt{\omega(k)}}{\sqrt{2}} \hat{\varphi}(k) x \cdot e(k,j).$$

We have

$$a(f,j)\Psi_\nu = -\int_{\mathbb{R}^3} f(k) \kappa_j^{\rm reg}(k) dk,$$

where $\mathscr{H}$-valued kernel is given by

$$\kappa_j^{\rm reg}(k) = (H_{\rm PF}^{\rm reg} - E + \omega(k))^{-1} C_j(k) \langle x \rangle \Psi_\nu.$$

Hence we can conclude the proposition below.

Proposition 3.22 $\Psi_\nu \in D(N^{1/2})$ *if and only if*

$$\sum_{j=1,2} \int_{\mathbb{R}^3} \|\kappa_j^{\rm reg}(k)\|_{\mathscr{H}}^2 dk < \infty, \tag{3.17}$$

and when (3.17) *holds it follows that*

$$\|N^{1/2}\Psi_\nu\|^2 = \sum_{j=1,2} \int_{\mathbb{R}^3} \|\kappa_j^{\rm reg}(k)\|_{\mathscr{H}}^2 dk.$$

This is called regularized pull-through formula. Suppose the binding condition. In Sect. 3.4.3 we show that there exists C such that $\sup_{\nu>0} \|e^{\beta|x|}\Psi_\nu\| < C$ for any normalized ground state Ψ_ν. Thus there exists a constant c such that $\sup_{\nu>0} \||x|\Psi_\nu\| \le c$.

Theorem 3.23 (Bach, Fröhlich and Sigal [9, Theorem 2.3]) *There exists a constant* C *independent of* ν *such that*

$$\|N^{1/2}\Psi_\nu\|^2 \le C\left(\int_{\mathbb{R}^3} \frac{|\hat{\varphi}(k)|^2}{\omega(k)} dk\right) \|\langle x\rangle \Psi_\nu\|^2, \quad \nu > 0. \tag{3.18}$$

In particular $\sup_{\nu>0} \|N^{1/2}\Psi_\nu\| < \infty$.

Proof We have

$$\begin{aligned}
\|N^{1/2}\Psi_\nu\|^2 &= \int_{\mathbb{R}^3} \left\|(H_{\rm PF}^{\rm reg} - E + \omega(k))^{-1} C_j(k)\langle x\rangle\Psi_\nu\right\|^2 dk \\
&\le \int_{\mathbb{R}^3} \frac{|\hat{\varphi}(k)|^2}{\omega(k)^3} \left\| e(k,j)(-i\nabla - A^R)\frac{(e^{-ikx}-1)}{\langle x\rangle}\langle x\rangle\Psi_\nu\right\|^2 dk + \int_{\mathbb{R}^3} \frac{|\hat{\varphi}(k)|^2}{\omega(k)} dk \|\Psi_\nu\|^2.
\end{aligned}$$

The first term on the right-hand side above can be estimated as

$$\begin{aligned}
&\int_{\mathbb{R}^3} \frac{|\hat{\varphi}(k)|^2}{\omega(k)^3} \left\| e(k,j)(-i\nabla - A^R)\frac{(e^{-ikx}-1)}{\langle x\rangle}\langle x\rangle\Psi_\nu\right\|^2 dk \\
&\le \int_{\mathbb{R}^3} \frac{|\hat{\varphi}(k)|^2}{\omega(k)^3} \left\| \frac{(e^{-ikx}-1)}{\langle x\rangle} e(k,j)(-i\nabla - A^R)\langle x\rangle\Psi_\nu\right\|^2 dk \\
&+ \int_{\mathbb{R}^3} \frac{|\hat{\varphi}(k)|^2}{\omega(k)^3} \left\| \frac{x}{\langle x\rangle^2} \frac{(e^{-ikx}-1)}{\langle x\rangle} e(k,j)\langle x\rangle\Psi_\nu\right\|^2 dk \\
&\le \int_{\mathbb{R}^3} \frac{|\hat{\varphi}(k)|^2}{\omega(k)} \|e(k,j)(-i\nabla - A^R)\langle x\rangle\Psi_\nu\|^2 dk + \int_{\mathbb{R}^3} \frac{|\hat{\varphi}(k)|^2}{\omega(k)} dk \|\Psi_\nu\|^2.
\end{aligned}$$

Note that

$$\sum_{j=1,2} \|e(k,j)(-i\nabla - A^R)\langle x\rangle\Psi_\nu\|^2 \le C_1(\langle x\rangle\Psi_\nu, H_{\rm PF}^{\rm reg}\langle x\rangle\Psi_\nu) + C_2\|\langle x\rangle\Psi_\nu\|^2$$
$$= (C_1E + C_2)\|\langle x\rangle\Psi_\nu\|^2 + C_1(\langle x\rangle\Psi_\nu, [H_{\rm PF}^{\rm reg}, \langle x\rangle]\Psi_\nu)$$
$$= (C_1E + C_2)\|\langle x\rangle\Psi_\nu\|^2 + C_1(\langle x\rangle\Psi_\nu, (2\frac{x}{\langle x\rangle}(-i\nabla - A) + \frac{1}{\langle x\rangle})\Psi_\nu).$$

Since $(\langle x\rangle\Psi_\nu, (2\frac{x}{\langle x\rangle}(-i\nabla - A) + \frac{1}{\langle x\rangle})\Psi_\nu) \le C_3\|\langle x\rangle\Psi_\nu\|^2$, we have

$$\sum_{j=1,2} \|e(k,j)(-i\nabla - A^R)\langle x\rangle\Psi_\nu\|^2 \le C_4\|\langle x\rangle\Psi_\nu\|^2$$

and (3.18) follows. ■

3.4.5 Derivative with Respect to Field Variable k

In this section we shall derive the weak derivative of $\Psi_\nu^{(n)}(x, k_1, \ldots, k_n)$ with respect to $k_{i,\mu}$, $i = 1, \ldots, n$ and $\mu = 1, 2, 3$. Throughout this section we assume that $\nu > 0$ and we also treat $H_{\rm PF}^{\rm reg}$ instead of $H_{\rm PF}$. We write the ground state of $H_{\rm PF}^{\rm reg}$ with positive mass ν as

$$\Psi_\nu = (\Psi_\nu^{(n)})_{n=0}^\infty \in \oplus_{n=0}^\infty \mathscr{H}^{(n)}$$

and we shall show that $\Psi_\nu^{(n)} \in W^{1,p}(S)$, i.e., $\nabla\Psi_\nu^{(n)} \in L^p(S)$ for $1 \le p < 2$ and $n \ge 1$ with some bounded open domain $S \subset \mathbb{R}_x^3 \times \mathbb{R}_k^{3n}$ with C^1 boundary. Note that $\Psi_\nu^{(0)} \in L^2(\mathbb{R}_x^3)$ and $\nabla_k\Psi_\nu^{(0)} = 0$. We see that

$$a(f,j)\Psi_\nu = -\int_{\mathbb{R}^3} f(k)\kappa^{\rm reg}(k)dk = -\int_{\mathbb{R}^3} f(k)R(k)C_j(k)\langle x\rangle\Psi_\nu dk.$$

Here we set $R(k) = (H_{\rm PF}^{\rm reg} - E + \omega(k))^{-1}$. We define

$$\Phi = \int_{\mathbb{R}^3}^\oplus R(k)C_j(k)\langle x\rangle\Psi_\nu dk, \quad \Phi^{(n+1)} = \int_{\mathbb{R}^3}^\oplus (R(k)C_j(k)\langle x\rangle\Psi_\nu)^{(n)}dk.$$

Note that

$$\Phi \in \int_{\mathbb{R}^3}^\oplus \mathscr{H}dk \cong L^2(\mathbb{R}_k^3)\otimes\mathscr{H}, \quad \Phi^{(n+1)} \in \int_{\mathbb{R}^3}^\oplus \mathscr{H}^{(n)}dk \cong \mathscr{H}^{(n+1)}.$$

Lemma 3.24 *Suppose $f \in L^2(\mathbb{R}_k^3)$ and $G \in \mathscr{H}^{(n)}$. Then*

$$(f \otimes G, \Psi_\nu^{(n+1)})_{\mathscr{H}^{(n+1)}} = -\frac{1}{\sqrt{n+1}}(f \otimes G, \Phi^{(n+1)})_{\mathscr{H}^{(n+1)}}.$$

Proof Suppose that $f, f/\sqrt{\omega} \in L^2(\mathbb{R}^3)$. We have

$$(\Psi, a(\bar{f}, j)\Psi_\nu)_{\mathscr{H}} = -\left(\Psi, \int_{\mathbb{R}^3} \bar{f}(k)R(k)C_j(k)\langle x\rangle\Psi_\nu dk\right)_{\mathscr{H}}$$

for any $\Psi \in \mathscr{H}$. Take $\Psi = (0, \ldots, 0, \overset{n_{th}}{G}, 0, \ldots) \in \mathscr{H}$, where $G \in \mathscr{H}^{(n)}$. Since

$$\begin{aligned}
&(\Psi, a(\bar{f}, j)\Psi_\nu^{(n+1)})_{\mathscr{H}^{(n)}} = (a^\dagger(f, j)\Psi, \Psi_\nu^{(n+1)})_{\mathscr{H}^{(n+1)}} \\
&= \sqrt{n+1}(S_{n+1}(f \otimes G), \Psi_{\mathrm{g}}^{(n+1)}(\nu))_{\mathscr{H}^{(n+1)}} \\
&= \sqrt{n+1}(f \otimes G, S_{n+1}\Psi_{\mathrm{g}}^{(n+1)}(\nu))_{\mathscr{H}^{(n+1)}} = \sqrt{n+1}(f \otimes G, \Psi_\nu^{(n+1)})_{\mathscr{H}^{(n+1)}}.
\end{aligned}$$

On the other hand we can see that

$$\begin{aligned}
&\left(\Psi, \int_{\mathbb{R}^3} \bar{f}(k)R(k)C_j(k)\langle x\rangle\Psi_\nu dk\right)_{\mathscr{H}} = \int_{\mathbb{R}^3} \bar{f}(k)\left(\Psi, R(k)C_j(k)\langle x\rangle\Psi_\nu\right)_{\mathscr{H}} dk \\
&= \int_{\mathbb{R}^3} \bar{f}(k)\left(G, (R(k)C_j(k)\langle x\rangle\Psi_\nu)^{(n)}dk\right)_{\mathscr{H}^{(n)}} = (f \otimes G, \Phi^{(n+1)})_{\mathscr{H}^{(n+1)}}.
\end{aligned}$$

Thus the lemma follows for f with $f, f/\sqrt{\omega} \in L^2(\mathbb{R}^3)$. By a limiting argument the lemma follows for all $f \in L^2(\mathbb{R}^3)$. ■

Corollary 3.25 *It follows that*

$$\Psi_\nu^{(n+1)}(x, k, k_1, \ldots, k_n) = -\frac{1}{\sqrt{n+1}}C_j(k)(R(k)\langle x\rangle\Psi_\nu)^{(n)}(x, k_1, \ldots, k_n). \quad (3.19)$$

In particular $\Psi_\nu^{(n+1)}\lceil_D = 0$, *where*

$$D = \mathbb{R}_x^3 \times \cup_{j=1}^{n+1} \underbrace{\mathbb{R}_{k_1}^3 \times \cdots \times \overset{j}{(\mathrm{supp}\hat{\varphi})^c} \times \cdots \times \mathbb{R}_{k_n}^3}_{n+1}. \quad (3.20)$$

Proof Let $f \in C_0^\infty(\mathbb{R}_k^3)$. By Lemma 3.24 we have

$$\begin{aligned}
&\int_{\mathbb{R}_k^3 \times \mathbb{R}_x^3 \times \mathbb{R}_{\mathbf{k}}^{3n}} f(k)G(x, \mathbf{k})\Psi_\nu^{(n+1)}(x, k, \mathbf{k})dxdkd\mathbf{k} \\
&= \int_{\mathbb{R}_k^3 \times \mathbb{R}_x^3 \times \mathbb{R}_{\mathbf{k}}^{3n}} f(k)G(x, \mathbf{k})C_j(k)(R(k)\langle x\rangle\Psi_\nu)^{(n)}(x, \mathbf{k})dxdkd\mathbf{k}.
\end{aligned}$$

Since $f \in C_0^\infty(\mathbb{R}^3)$ and $G \in \mathscr{H}^{(n)}$, the set of linear combinations of the form $f(k)G(x, \mathbf{k})$ is dense in $\mathscr{H}^{(n+1)}$, (3.19) follows. $\Psi_\nu^{(n+1)}\lceil_D = 0$ follows from the fact that $C_j(k) = \hat{\varphi}(k) \times$ operator and $C_j(k)$ is the permutation invariance of $(k, k_1, \ldots, k_n)$. ■

By Corollary 3.25 we can consider the weak derivative of $\Psi_\nu^{(n+1)}(x, k_1, \ldots, k_{n+1})$ with respect to $k_{i,\mu}$ for $i = 1, ..., n$ and $\mu = 1, 2, 3$. Let $\varepsilon_1 = (\varepsilon, 0, 0)$, $\varepsilon_2 = (0, \varepsilon, 0)$ and $\varepsilon_3 = (0, 0, \varepsilon)$. For $f \in C_0^\infty(\mathbb{R}^3)$ we can see by Lemma 3.24 that

$$\begin{aligned}(\nabla_{k_\mu} f \otimes G, \Psi_\nu^{(n+1)})_{\mathscr{H}^{(n+1)}} &= \lim_{\varepsilon \to 0} \int_{\mathbb{R}^3} \frac{\bar{f}(k + \varepsilon_\mu) - \bar{f}(k)}{\varepsilon} (G, \left(R(k)C_j(k)\langle x\rangle\Psi_\nu\right)^{(n)}) dk \\ &= \frac{1}{\sqrt{n+1}} \lim_{\varepsilon \to 0} (f \otimes G, \Phi_{\varepsilon_\mu}^{(n+1)})_{\mathscr{H}^{(n+1)}},\end{aligned}$$

where $\mu = 1, 2, 3$, and $\Phi_{\varepsilon_\mu}^{(n+1)} = \int_{\mathbb{R}^3}^\oplus \Phi_{\varepsilon_\mu}^{(n)}(k) dk$ with

$$\Phi_{\varepsilon_\mu}^{(n)}(k) = \left(\frac{R(k + \varepsilon_\mu)C_j(k + \varepsilon_\mu) - R(k)C_j(k)}{\varepsilon} \langle x\rangle\Psi_\nu\right)^{(n)}.$$

We fix $1 \le \mu \le 3$. We shall investigate the convergence of $(\Phi_{\varepsilon_\mu}^{(n+1)})_{\varepsilon_\mu \in \mathbb{R}\setminus\{0\}}$ as $\varepsilon \to 0$. Let

$$\Phi_\mu(k) = (R_\mu(k)C_j(k) + R(k)C_j^\mu(k))\langle x\rangle\Psi_\nu,$$

where

$$\begin{aligned}R_\mu(k) &= R(k)(\nabla_\mu\omega(k))R(k), \\ C_j^\mu(k) &= \left(\nabla_\mu \frac{\hat{\varphi}}{\sqrt{\omega}}\right)(k)P_j(k) + \frac{\hat{\varphi}(k)}{\sqrt{\omega(k)}}(\nabla_\mu P_j)(k) + (\nabla_\mu \rho_j)(k)\frac{1}{\langle x\rangle}.\end{aligned}$$

Here $P_j(k) = e(k, j)(e^{-ikx} - 1)(-i\nabla - A^R)$. By the definition of Φ we keep in mind that

$$-\frac{1}{\sqrt{n+1}}\Phi_\mu^{(n+1)}(k, k_1, \ldots, k_n) = \nabla_{k_\mu}\Psi_\nu^{(n+1)}(k, k_1, \ldots, k_n).$$

We shall estimate

$$\begin{aligned}&\Phi_{\varepsilon_\mu}(k) - \Phi_\mu(k) \\ &= \left(\frac{R(k + \varepsilon_\mu)C_j(k + \varepsilon_\mu) - R(k)C_j(k)}{\varepsilon} - R_\mu(k)C_j(k) - R(k)C_j^\mu(k)\right)\langle x\rangle\Psi_\nu.\end{aligned} \tag{3.21}$$

We divide (3.21) as

$$\Phi_{\varepsilon_\mu}(k) - \Phi_\mu(k) = \sum_{i=1}^{8} G_i(k),$$

where

$$\begin{aligned}
G_1(k) &= \left(\frac{R(k+\varepsilon_\mu) - R(k)}{\varepsilon} - R_\mu(k)\right) C_j(k+\varepsilon_\mu)\langle x\rangle \Psi_\nu,\\
G_2(k) &= R_\mu(k)(C_j(k+\varepsilon_\mu) - C_j(k))\langle x\rangle \Psi_\nu,\\
G_3(k) &= R(k)\left(\frac{\rho_j(k+\varepsilon_\mu) - \rho_j(k)}{\varepsilon} - \rho_j^\mu(k)\right)\Psi_\nu,\\
G_4(k) &= R(k)\left(\frac{\hat{\varphi}_\omega(k+\varepsilon_\mu) - \hat{\varphi}_\omega(k)}{\varepsilon} - \hat{\varphi}_\omega^\mu(k)\right) P_j(k+\varepsilon_\mu)\langle x\rangle \Psi_\nu,\\
G_5(k) &= R(k)\hat{\varphi}_\omega(k)(P_j(k+\varepsilon_\mu) - P_j(k))\langle x\rangle \Psi_\nu,\\
G_6(k) &= R(k)\hat{\varphi}_\omega(k)\left(\frac{T_j(k+\varepsilon_\mu) - T_j(k)}{\varepsilon} - T_j^\mu(k)\right)\eta(k+\varepsilon_\mu)\Psi_\nu,\\
G_7(k) &= R(k)\hat{\varphi}_\omega(k)T_j^\mu(k)(\eta(k+\varepsilon_\mu) - \eta(k))\Psi_\nu,\\
G_8(k) &= R(k)\hat{\varphi}_\omega(k)T_j(k)\left(\frac{\eta(k+\varepsilon_\mu) - \eta(k)}{\varepsilon} - \eta^\mu(k)\right)\Psi_\nu.
\end{aligned}$$

Here we set $\hat{\varphi}_\omega = \hat{\varphi}/\sqrt{\omega}$, $\nabla_\mu \hat{\varphi}_\omega = \hat{\varphi}_\omega^\mu$, $\nabla_\mu \rho_j = \rho_j^\mu$, $T_j(k) = e(k,j)(-i\nabla - A^R)$, $T_j^\mu(k) = (\nabla_\mu e(k,j))(-i\nabla - A^R)$, $\eta(k) = e^{-ikx} - 1$ and $\eta^\mu(k) = -ix_\mu e^{-ikx}$. In the definitions of $T_j^\mu(k)$ and $\rho_j^\mu(k)$, the partial derivative of $e(k,j)$, $\nabla_\mu e(k,j)$, appears. Since polarization vectors are given by (3.1), $e_\mu(k,j) \in C^\infty(\mathbb{R}^3 \setminus K_{\mathrm{pol}})$ for $\mu = 1,2,3$ and $j = 1,2$, where

$$K_{\mathrm{pol}} = \{(k_1,k_2,k_3) \in \mathbb{R}^3 | k_1^2 + k_2^2 = 0\}.$$

There exists a constant C such that for $k \in \mathbb{R}^3 \setminus K_{\mathrm{pol}}$,

$$|\nabla_\mu e(k,j)| \le \frac{C}{\sqrt{k_1^2+k_2^2}}, \quad |\nabla_\mu^2 e(k,j)| \le \frac{C}{k_1^2+k_2^2}.$$

We have the lemma.

Lemma 3.26 *There exists a constant C such that*

$$\|C_j(k)\langle x\rangle \Psi_\nu\| \le C\sqrt{\omega(k)}|\hat{\varphi}(k)|\|\langle x\rangle \Psi_\nu\|.$$

Proof We see that

$$\|C_j(k)\Psi_\nu\| \le \frac{|\hat\varphi(k)|}{\sqrt{2\omega(k)}}\|(e^{-ikx}-1)e(k,j)(-i\nabla - A)\frac{1}{\langle x\rangle}\Psi_\nu\| + \frac{\sqrt{\omega(k)}}{\sqrt{2}}|\hat\varphi(k)|\|\Psi_\nu\|$$
$$\le \frac{|\hat\varphi(k)|}{\sqrt{2\omega(k)}}\|\frac{e^{-ikx}-1}{\langle x\rangle}e(k,j)(-i\nabla - A)\Psi_\nu\| + \frac{|\hat\varphi(k)|}{\sqrt{2\omega(k)}}\|\frac{e^{-ikx}-1}{\langle x\rangle}\frac{ixe(k,j)}{\langle x\rangle^2}\Psi_\nu\| + \frac{\sqrt{\omega(k)}}{\sqrt{2}}|\hat\varphi(k)|\|\Psi_\nu\|$$
$$\le \frac{\sqrt{\omega(k)}|\hat\varphi(k)|}{\sqrt{2}}\left(\|e(k,j)(-i\nabla - A)\Psi_\nu\| + \|\frac{1}{\langle x\rangle}\Psi_\nu\| + \|\Psi_\nu\|\right).$$

Then

$$\|C_j(k)\langle x\rangle\Psi_\nu\| \le \frac{\sqrt{\omega(k)}|\hat\varphi(k)|}{\sqrt{2}}\left(\|e(k,j)(-i\nabla - A)\langle x\rangle\Psi_\nu\| + \|\Psi_\nu\| + \|\langle x\rangle\Psi_\nu\|\right).$$

Since $\|e(k,j)(-i\nabla - A)\langle x\rangle\Psi_\nu\| \le C\|\langle x\rangle\Psi_\nu\|$ with some constant C, we prove the lemma. ■

Next we estimate $\|(\nabla_\mu C_j)(k)\Psi_\nu\|$.

Lemma 3.27 *Suppose that $\hat\varphi \in C^1(\mathbb{R}^3)$. Then there exists a constant C such that*

$$\|(\nabla_\mu C_j)(k)\langle x\rangle\Psi_\nu\| \le C\xi(k)\|\langle x\rangle\Psi_\nu\|, \quad k \in \mathbb{R}^3 \setminus K_{\rm pol},$$

where $\xi(k) = \sqrt{\omega(k)}\left(|\nabla_\mu\hat\varphi(k)| + \frac{|\hat\varphi(k)|}{\sqrt{k_1^2+k_2^2}}\right)$.

Proof Set $\partial = \nabla_\mu$. We have

$$|\partial\frac{\hat\varphi}{\sqrt{\omega}}| \le C(\frac{|\partial\hat\varphi|}{\sqrt{\omega(k)}} + \frac{|\hat\varphi|}{\omega(k)^{3/2}}),\ |\partial(\hat\varphi\sqrt{\omega})| \le C(|\partial\hat\varphi|\sqrt{\omega} + \frac{|\hat\varphi|}{\sqrt{\omega(k)}}).$$

Hence we have $\partial C_j(k) = \sum_{i=1}^5 R_i$, where

$$R_1 = (\partial\frac{\hat\varphi}{\sqrt{2\omega}})e(k,j)(-i\nabla - A^R)\frac{1-e^{-ikx}}{\langle x\rangle}, \quad R_2 = \frac{\hat\varphi}{\sqrt{2\omega}}(\partial e(k,j))(-i\nabla - A^R)\frac{1-e^{-ikx}}{\langle x\rangle},$$
$$R_3 = \frac{\hat\varphi}{\sqrt{2\omega}}e(k,j)(-i\nabla - A^R)\frac{-ix_\mu}{\langle x\rangle}, \quad R_4 = -i\partial(\sqrt{\omega}\hat\varphi)\frac{e(k,j)x}{\langle x\rangle}, \quad R_5 = -i\sqrt{\omega}\hat\varphi(\partial e(k,j))\frac{x}{\langle x\rangle}.$$

We see that

$$\|R_1\langle x\rangle\Psi_\nu\| \le C(\frac{|\partial\hat\varphi(k)|}{\sqrt{\omega(k)}} + \frac{|\hat\varphi(k)|}{\omega(k)^{3/2}})\|(1-e^{-ikx})e(k,j)(-i\nabla - A^R)\Psi_\nu\|$$
$$\le C'(|\partial\hat\varphi(k)|\sqrt{\omega(k)} + \frac{|\hat\varphi(k)|}{\sqrt{\omega(k)}})\|\langle x\rangle\Psi_\nu\|,$$
$$\|R_2\langle x\rangle\Psi_\nu\| \le C\frac{1}{\sqrt{\omega(k)}}\frac{|\hat\varphi(k)|}{\sqrt{k_1^2+k_2^2}}\|(1-e^{-ikx})(-i\nabla - A^R)\Psi_\nu\|$$
$$\le C'\sqrt{\omega(k)}\frac{|\hat\varphi(k)|}{\sqrt{k_1^2+k_2^2}}\|\langle x\rangle\Psi_\nu\|,$$

$$\|R_3\langle x\rangle\Psi_\nu\| \le C\frac{|\hat{\varphi}(k)|}{\sqrt{\omega(k)}}(\|H_{\rm PF}^{\rm reg}x_\mu\Psi_\nu\| + \|x_\mu\Psi_\nu\|)$$
$$\le C'\frac{|\hat{\varphi}(k)|}{\sqrt{\omega(k)}}(\|\langle x\rangle\Psi_\nu\| + \|\Psi_\nu\|) \le C''\frac{|\hat{\varphi}(k)|}{\sqrt{k_1^2+k_2^2}}\|\langle x\rangle\Psi_\nu\|,$$

$$\|R_4\langle x\rangle\Psi_\nu\| \le C(\sqrt{\omega(k)}|\partial\hat{\varphi}(k)| + \frac{|\hat{\varphi}(k)|}{\sqrt{\omega(k)}})\|\langle x\rangle\Psi_\nu\|,$$

$$\|R_5\langle x\rangle\Psi_\nu\| \le C\sqrt{\omega(k)}\frac{|\hat{\varphi}(k)|}{\sqrt{k_1^2+k_2^2}}\|\langle x\rangle\Psi_\nu\|.$$

Hence we complete the proof. ■

Corollary 3.28 (Griesemer, Lieb and Loss [21, Theorem 6.3]) *It follows that*

$$\|\Phi_\mu(k)\| \le C\left(\frac{|\hat{\varphi}(k)|}{\omega(k)^{3/2}} + \frac{|\partial\hat{\varphi}(k)|}{\sqrt{\omega(k)}} + \frac{|\hat{\varphi}(k)|}{\sqrt{\omega(k)}\sqrt{k_1^2+k_2^2}}\right)\|\langle x\rangle\Psi_\nu\|.$$

Proof By the definition of $\Phi_\mu(k)$ and the estimates given by Lemmas 3.26 and 3.27, the corollary follows. ■

We shall estimate $G_1, \ldots, G_8$ in the following lemma. We note that in the case of $\nu > 0$ there exist $0 < c \le C$ independent of $k \in \mathbb{R}^3$ and $-1 \le \varepsilon \le 1$ such that

$$c \le \frac{\omega(k+\varepsilon_\mu)}{\omega(k)} \le C.$$

Lemma 3.29 *Suppose that $\hat{\varphi} \in C_0^2(\mathbb{R}^3)$. Let $\phi \in C_0^\infty(\mathbb{R}^3 \setminus K_{\rm pol})$. It follows that*

$$\lim_{\varepsilon\to 0}\int_{\mathbb{R}^3}\phi(k)\|G_i(k)\|_{\mathscr{H}}dk = 0, \quad i = 1, \ldots, 8.$$

Proof Suppose that ${\rm supp}\hat{\varphi} \subset S$. We can see that on $S \setminus K_{\rm pol}$, $\|G_i(k)\| \le |\varepsilon|C\|\langle x\rangle\Psi_\nu\|$ for $i = 1, 2, 3, 4, 7$, and $\|G_i(k)\| \le |\varepsilon|C\|\langle x\rangle^2\Psi_\nu\|$ for $i = 8$, and G_5 and G_6 can be estimated as $\|G_5(k)\| \le |\varepsilon|C\left(\frac{1}{\sqrt{k_1^2+k_2^2}} + 1\right)\|\langle x\rangle\Psi_\nu\|$ and $\|G_6\| \le |\varepsilon|\frac{C}{k_1^2+k_2^2}\|\langle x\rangle\Psi_\nu\|$. Then the proof is complete. ■

Lemma 3.30 *Let $f \in C_0^\infty(\mathbb{R}^3)$ and $G \in \mathscr{H}^{(n)}$. Then it follows that*

$$\lim_{\varepsilon\to 0}(f\otimes G, \Phi_{\varepsilon_\mu}^{(n+1)}) = (f\otimes G, \Phi_\mu^{(n+1)}).$$

In particular it follows that

$$(\nabla_\mu f \otimes G, \Psi_\nu^{(n+1)}) = \frac{1}{\sqrt{n+1}}(f \otimes G, \Phi_\mu^{(n+1)}).$$

Proof It follows that $(\nabla_\mu f \otimes G, \Psi_\nu^{(n+1)}) = \lim_{j\to\infty}(\nabla_\mu f_j \otimes G, \Psi_\nu^{(n+1)})$, where $f_j \in C_0^\infty(\mathbb{R}^3 \setminus K_{\rm pol})$ and $\nabla_\mu f_j \to \nabla_\mu f$ in $L^2(\mathbb{R}_k^3)$. Note that $\Phi_{\varepsilon_\mu}(k) - \Phi_\mu(k) = \sum_{i=1}^8 G_i(k)$ and then

$$\left|\int_{\mathbb{R}^3} f_j(k)(G, \Phi_{\varepsilon_\mu}^{(n)}(k) - \Phi_\mu^{(n)}(k))_{\mathscr{H}^{(n)}} dk\right| \le \|G\| \sum_{i=1}^8 \int_{\mathbb{R}^3} |f_j(k)| \|G_i(k)\| dk \to 0$$

as $\varepsilon \to 0$ by Lemmas 3.29. For $f \in C_0^\infty(\mathbb{R}^3 \setminus K_{\rm pol})$, we have

$$\begin{aligned}
(\nabla_\mu f \otimes G, \Psi_\nu^{(n+1)})_{\mathscr{H}^{(n+1)}} &= \frac{1}{\sqrt{n+1}} \lim_{\varepsilon\to 0}(f \otimes G, \Phi_{\varepsilon_\mu}^{(n+1)})_{\mathscr{H}^{(n+1)}} \\
&= \frac{1}{\sqrt{n+1}} \lim_{\varepsilon\to 0} \int_{\mathbb{R}^3} f(k)(G, \Phi_{\varepsilon_\mu}^{(n)}(k))_{\mathscr{H}^{(n)}} dk = \frac{1}{\sqrt{n+1}} \int_{\mathbb{R}^3} f(k)(G, \Phi_\mu^{(n)}(k))_{\mathscr{H}^{(n)}} dk \\
&= \frac{1}{\sqrt{n+1}}(f \otimes G, \Phi_\mu^{(n+1)})
\end{aligned}$$

and the proof is complete. ■

Lemma 3.31 (Derivative with respect to field variable k) *Suppose $\hat{\varphi} \in C_0^2(\mathbb{R}^3)$. Let $S = D^c \subset \mathbb{R}_x^3 \times \mathbb{R}_k^{3n}$, where D is given by (3.20) with dimension n replaced by $n+1$. Then for arbitrary $n \ge 1$, $\Psi_\nu^{(n)}$ is weakly differentiable with respect to $k_{i,\mu}$, $i = 1, ..., n$, $\mu = 1, 2, 3$. Moreover if $1 \le p < 2$, then*

$$\sup_{0<\nu<1} \|\nabla_{k_{i,\mu}} \Psi_\nu^{(n)}\|_{L^p(S)} < \infty. \tag{3.22}$$

Proof Let $n \ge 0$. By Lemma 3.30 we can conclude that

$$\nabla_\mu \Psi_\nu^{(n+1)} = \frac{1}{\sqrt{n+1}} \Phi_\mu^{(n+1)}.$$

We then see that $\|\nabla_\mu \Psi_\nu^{(n+1)}\|_{L^p(S)}^p \le C \int_{\mathbb{R}^3} \|\Phi_\mu^{(n)}(k)\|_{\mathscr{H}^{(n)}}^p dk$. By Corollary 3.28 we have

$$\int_{\mathbb{R}^3} \|\Phi_\mu^{(n)}(k)\|_{\mathscr{H}^{(n)}}^p dk \le C^p \int_{\mathbb{R}^3} \left| \frac{|\hat{\varphi}(k)|}{\omega(k)^{3/2}} + \frac{|\partial\hat{\varphi}(k)|}{\sqrt{\omega(k)}} + \frac{|\hat{\varphi}(k)|}{\sqrt{\omega(k)}\sqrt{k_1^2 + k_2^2}} \right|^p dk \|\langle x\rangle \Psi_\nu\|^p < \infty$$

for $1 \le p < 2$. Furthermore $\sup_{\nu>0} \|\langle x\rangle \Psi_\nu\|^2 < \infty$. Thus (3.22) follows. ■

3.4.6 Derivative with Respect to Matter Variable *x*

We consider the derivative of $\Psi_\nu^{(n)}(x, k_1, ..., k_n)$ with respect to x. Let E_ν be the ground state energy of $H_{\rm PF}$ with artificial mass ν and recall that E is that for $\nu = 0$.

Lemma 3.32 *It follows that* $\lim_{\nu\to 0} E_\nu = E$.

Proof Since E_ν is monotonously non-increasing with respect to ν and bounded from below, there exists E^* such that $\lim_{\nu\to 0} E_\nu = E^*$ and $E \le E^*$. We shall show that $E \ge E^*$. Let $\Phi \in Q(\mathrm{q})$ be such that $\mathrm{q}(\Phi, \Phi) < E + \varepsilon$. Let P_n be the projection to subspace $\oplus_{m=0}^n \mathscr{F}^{(m)}$, and set $\tilde{P}_n = 1\!\!1_{L_2} \otimes P_n$. We can see that $\mathrm{q}(\tilde{P}_n\Phi, \tilde{P}_n\Phi)/\|\tilde{P}_n\Phi\|^2 \to \mathrm{q}(\Phi, \Phi)$ as $n \to \infty$. Hence there exists $\Phi \in Q(\mathrm{q}) \cap D(N)$ such that $\mathrm{q}(\Phi, \Phi) < E + 2\varepsilon$. We have

$$E^* \le E_\nu \le \mathrm{q}(\Phi, \Phi) + \nu(\Phi, N\Phi) \le E + 2\varepsilon + \nu(\Phi, N\Phi).$$

Here we used $\mathrm{d}\Gamma(\sqrt{|k|^2 + \nu^2}) \le \mathrm{d}\Gamma(|k|) + \nu N$. Taking the limit of $\nu \to 0$ on both sides we have $E^* \le E + 2\varepsilon$, which implies that $E^* \le E$. Thus $\lim_{\nu\to 0} E_\nu = E$ follows. ■

Lemma 3.33 (Derivative with respect to matter variable x) *Suppose that* $\mathrm{supp}\hat{\varphi}$ *is compact. Then for arbitrary* $n \ge 0$, $\Psi_\nu^{(n)} = \Psi_\nu^{(n)}(x, k_1, ..., k_n)$ *is differentiable with respect to* x_μ, $\mu = 1, 2, 3$. *Moreover if* $1 \le p < 2$ *and* $R > 0$. *Then*

$$\sup_{0<\nu<1} \|1\!\!1_{|x|\le R}\nabla_{x_\mu}\Psi_\nu^{(n)}\|_{L^p(\mathbb{R}_x^3\times\mathbb{R}_k^{3n})} < \infty.$$

Proof We have $\|\nabla_{x_\mu}\Psi_\nu\|_{\mathscr{H}} \le C\|(H_{\rm PF} + 1\!\!1)\Psi_\nu\|_{\mathscr{H}} \le C|E_\nu + 1| \le C'$. Here C' can be chosen independently of ν by Lemma 3.32. Thus $\|\nabla_{x_\mu}\Psi_\nu\|_{\mathscr{H}} \le C'$ follows and $\|\nabla_{x_\mu}\Psi_\nu^{(n)}\|_{\mathscr{H}^{(n)}} \le C'$. Since $\hat{\varphi}$ has a compact support, $\nabla_{x_\mu}\Psi_\nu^{(n)}\lceil_D = 0$, where D is in (3.20) with n replaced by $n + 1$, and D^c is compact. By using the Hölder inequality twice we have

$$\begin{aligned}
&\|1\!\!1_{|x|\le R}\nabla_{x_\mu}\Psi_\nu^{(n)}\|^p_{L^p(\mathbb{R}_x^3\times\mathbb{R}_k^{3n})} = \int_{\mathbb{R}_k^{3n}} dk \int_{\mathbb{R}_x^3} 1\!\!1_{|x|\le R}|\nabla_{x_\mu}\Psi_\nu^{(n)}(x,k)|^p dx \\
&\le \int_{\mathbb{R}_k^{3n}} dk \left(\int_{\mathbb{R}_x^3} 1\!\!1_{|x|\le R} dx\right)^{\frac{2-p}{2}} \left(\int_{\mathbb{R}_x^3} |\nabla_{x_\mu}\Psi_\nu^{(n)}(x,k)|^2 dx\right)^{\frac{p}{2}} \\
&\le C\int_{D^c} dk \left(\int_{\mathbb{R}_x^3} |\nabla_{x_\mu}\Psi_\nu^{(n)}(x,k)|^2 dx\right)^{\frac{p}{2}} \le C''\left(\int_{\mathbb{R}_x^3\times\mathbb{R}_k^{3n}} |\nabla_{x_\mu}\Psi_\nu^{(n)}(x,k)|^2 dxdk\right)^{\frac{p}{2}} \\
&= C''\|\nabla_{x_\mu}\Psi_\nu^{(n)}\|^p_{\mathscr{H}^{(n+1)}} < \infty.
\end{aligned}$$

Then the lemma follows. ■

3.4.7 Existence of the Ground State

The strategy to show the existence of the ground state is to apply a compact embedding of Sobolev spaces. Let Ψ_ν be the normalized ground state of $H_{\rm PF}^{\rm reg}$ with artificial positive mass ν. Then there exists a sequence Ψ_{ν_l} such that Ψ_{ν_l} weakly converges to Φ. It is sufficient to show that $\Phi \neq 0$ by the proposition below:

Proposition 3.34 (Arai and Hirokawa [5, Lemma 4.9]) *Let S_n and S be selfadjoint operators on a Hilbert space $\mathscr{K}$, which have a common core D such that $S_n \to S$ on D strongly as $n \to \infty$. Let ψ_n be a normalized eigenvector of S_n such that $S_n\psi_n = E_n\psi_n$, $E = \lim_{n\to\infty} E_n$ and the weak limit $\psi = w - \lim_{n\to\infty}\psi_n \neq 0$ exist. Then $S\psi = E\psi$. In particular if E_n is the ground state energy, then E is the ground state energy of S and ψ is a ground state of S.*

Proof Since S_n converges to S in the strong resolvent sense by the assumption, we can see that $\lim_{n\to\infty}(\phi, (S_n - z)^{-1}\psi_n) = (\phi, (S - z)^{-1}\psi)$ for any $\phi \in \mathscr{K}$. This implies that $(S_n - z)^{-1}\psi = (E - z)^{-1}\psi$ and then $S\psi = E\psi$. ■

Theorem 3.35 (Griesemer, Lieb and Loss [21, Theorem 2.1]) *Let $\nu = 0$. Suppose Assumption 3.10, $\hat{\varphi} \in C_0^2(\mathbb{R}^3)$, and $V \in V_{\rm Kato} \cup V_{\rm Kato}^{\rm q}$. Then $H_{\rm PF}^{\rm reg}$ has the ground state. In particular $H_{\rm PF}$ also has the ground state.*

Proof We suppose that $\|\Psi_\nu\| = 1$. Then there exists a sequence $\{\Psi_{\nu_l}\}_{l=1}^\infty$ such that $\lim_{l\to\infty}\nu_l = 0$ and $w-\lim \Psi_{\nu_l} = \Phi$. Set $\Phi_l = \Psi_{\nu_l}$ for simplicity. By Proposition 3.34 it is sufficient to show that $\Phi \neq 0$. Suppose that $\operatorname{supp}\hat{\varphi} \subset B_r$, where $B_r \subset \mathbb{R}^3$ denotes the open ball of radius r centered at the origin in $\mathbb{R}^3$. Hence $\Psi_\nu^{(n)}\lceil_{\mathbb{R}_x^3\times(\underbrace{B_r\times\cdots\times B_r}_{n})^c} = 0$. Let $B_R^n \subset \mathbb{R}^{3+3n}$ be an arbitrary open ball such that

$$B_R^n \supset \{(x, k_1, \ldots, k_n) \in \mathbb{R}_x^3 \times \mathbb{R}_k^{3n} | x \in B_R, k_j \in B_r, j = 1, ..., n\}.$$

We have

$$\|\Phi_l - \Phi_m\|_{\mathscr{H}}^2 \le \sum_{n=0}^M \|\Phi_l^{(n)} - \Phi_m^{(n)}\|_{\mathscr{H}^{(n)}}^2 + \sum_{n=M+1}^\infty \|\Phi_l^{(n)} - \Phi_m^{(n)}\|_{\mathscr{H}^{(n)}}^2.$$

Since the support of $1\!\!1_{\{|x|<R\}}(\Phi_l^{(n)} - \Phi_m^{(n)})$ is contained in B_R^n, the first term on the right-hand side is estimated as

$$\begin{aligned}\|\Phi_l^{(n)} - \Phi_m^{(n)}\|_{\mathscr{H}^{(n)}}^2 &= \|1\!\!1_{\{|x|<R\}}(\Phi_l^{(n)} - \Phi_m^{(n)})\|^2 + \|1\!\!1_{\{|x|\ge R\}}(\Phi_l^{(n)} - \Phi_m^{(n)})\|^2 \\ &\le \|1\!\!1_{\{|x|<R\}}(\Phi_l^{(n)} - \Phi_m^{(n)})\|_{L^2(B_R^n)}^2 + \frac{1}{R}\||x|(\Phi_l^{(n)} - \Phi_m^{(n)})\|^2 \\ &\le \|1\!\!1_{\{|x|<R\}}(\Phi_l^{(n)} - \Phi_m^{(n)})\|_{L^2(B_R^n)}^2 + \frac{2C_1}{R}\end{aligned}$$

for $0 \le n \le M$, where $C_1 = \max_{n\in\mathbb{N}} \sup_{l\in\mathbb{N}} \||x|\Phi_l^{(n)}\|^2$. The second term is

$$\sum_{n=M+1}^{\infty} \|\Phi_l^{(n)} - \Phi_m^{(n)}\|^2_{\mathscr{H}^{(n)}} \le \frac{1}{M}(\|N^{1/2}\Phi_l\| + \|N^{1/2}\Phi_m\|) \le \frac{2C_2}{M}.$$

Here we used $N/M \ge 1$ on $\mathscr{H}^{(n)}$ for $n \ge M$, and $C_2 = \sup_{l\in\mathbb{N}} \|N^{1/2}\Phi_l\|^2$. Together with them we obtain that

$$\|\Phi_l - \Phi_m\|^2_{\mathscr{H}} < \sum_{n=0}^{M} \|\Phi_l^{(n)} - \Phi_m^{(n)}\|^2_{L^2(B_R^n)} + \frac{2C_2}{M} + \frac{2C_1}{R}. \quad (3.23)$$

By Lemmas 3.31 and 3.33, $(\|\Phi_l^{(n)}\|_{W^{1,p}(B_R^n)})_{l=1}^{\infty}$ is a bounded sequence for each $p \in (1, 2)$, and B_R^n is an open ball, i.e., it is a bounded open set and the boundary is C^1 class. By the Rellich–Kondrachov compactness theorem,

$$W^{1,p}(B_R^n) \subset\subset L^q(B_R^n)$$

for any $1 \le q < \frac{p(3+3n)}{3+3n-p}$. Then as $l \to \infty$, $\Phi_l^{(n)}$ strongly converges to $\Phi^{(n)}$ in $L^q(B_R^n)$. In particular, taking $p > \frac{6}{5}$, we have

$$s-\lim_{n\to\infty} \Phi_l^{(n)} = \Phi^{(n)}, \quad 0 \le n \le M$$

in $L^2(B_R^n)$. Thus (3.23) implies that $\{\Phi_l\}_{l=1}^{\infty}$ is a Cauchy sequence in $\mathscr{H}$ and we can see that $s-\lim_{l\to 0} \Phi_l = \Phi$. Hence $\|\Phi\| = 1$ and in particular $\Phi \ne 0$. Then the proof is complete. ■

Chapter 4
The Nelson Model

Abstract In this chapter the Nelson model is studied, which was investigated by Nelson [49] who succeeded in renormalizing an ultraviolet cutoff function and defining Hamiltonian without cutoffs. This model describes an interaction between non-relativistic spinless nucleons and a scalar meson field. The existence of the ground state of the Nelson Hamiltonian with/without cutoffs is proven.

4.1 The Nelson Hamiltonian

4.1.1 Definition of the Nelson Hamiltonian

We consider a system of quantum matters linearly coupled to massless and/or massive bosons, and suppose that the space dimension is d for mathematical generality. This model describes, physically, an interaction between non-relativistic spinless nucleons and scalar mesons in an atomic nucleus. Then the state space of the N-quantum matters is described by $L^2(\mathbb{R}^{dN})$ and that of bosons by the boson Fock space over $L^2(\mathbb{R}^d)$; $\mathscr{F} = \mathscr{F}(L^2(\mathbb{R}^d))$. For simplicity we set $N = 1$. We also assume that the boson mass is zero. The total Hilbert space of the system is given by

$$\mathscr{H} = L^2(\mathbb{R}^d) \otimes \mathscr{F}.$$

It can be decomposed into infinite direct sum:

$$\mathscr{H} = \oplus_{n=0}^{\infty} \mathscr{H}^{(n)},$$

where $\mathscr{H}^{(n)} = L^2(\mathbb{R}^d) \otimes \mathscr{F}^{(n)}$. Let $\omega = \omega(k) = |k|$ be the dispersion relation which presents an energy of a single massless boson with momentum $k \in \mathbb{R}^d$. The free field Hamiltonian of the Nelson model is then given by

$$H_{\mathrm{f}} = \mathrm{d}\Gamma(\omega).$$

The quantum matter is governed by the Schrödinger operator H_{p}.

F. Hiroshima, *Ground States of Quantum Field Models*,
SpringerBriefs in Mathematical Physics,
https://doi.org/10.1007/978-981-32-9305-2_4

Assumption 4.1 (*External potential*) V satisfies that (1) $V = V_+ - V_- \in V_{\text{Kato}}^{\text{q}}$. (2) H_{p} is a non-negative selfadjoint operator and has a compact resolvent.

H_{p} is given by

$$H_{\text{p}} = -\frac{1}{2}\Delta \dotplus V_+ \dot{-} V_-,$$

where $\dotplus$ denotes the quadratic form sum.

To introduce a linear interaction let us now define a field operator ϕ in $\mathscr{H}$. We introduce assumptions on cutoff function $\hat{\varphi}$ which are assumed throughout this chapter unless otherwise stated.

Assumption 4.2 (*Cutoff functions*) $\varphi \in \mathscr{S}'(\mathbb{R}^d)$ satisfies that (1) $\hat{\varphi} \in L^1_{\text{loc}}(\mathbb{R}^d)$, (2) $\hat{\varphi}(-k) = \overline{\hat{\varphi}(k)}$, (3) $\hat{\varphi}/\sqrt{\omega}, \hat{\varphi}/\omega \in L^2(\mathbb{R}^d)$.

Let

$$\rho(x) = \rho(x, k) = \hat{\varphi}(k)e^{-ikx}/\sqrt{\omega(k)}, \quad x \in \mathbb{R}^d.$$

For each $x \in \mathbb{R}^d$, $\rho(x, \cdot) \in L^2(\mathbb{R}^d)$ and we set

$$\phi(\rho(x)) = \frac{1}{\sqrt{2}}\left(a^\dagger(\rho(x)) + a(\tilde{\rho}(x))\right),$$

where $\tilde{\rho}(x, k) = \rho(x, -k)$. Since $\overline{\hat{\varphi}(k)} = \hat{\varphi}(-k)$, $\phi(\rho(x))$ is essentially selfadjoint for each $x \in \mathbb{R}^d$ on $\mathscr{F}_{\text{fin}}$ and we denote $\overline{\phi(\rho(x))\lceil_{\mathscr{F}_{\text{fin}}}}$ by the same symbol $\phi(\rho(x))$. The field operator ϕ is defined by the constant fiber direct integral of selfadjoint operator $\phi(\rho(x))$:

$$\phi = \phi(\rho) = \int_{\mathbb{R}^d}^{\oplus} \phi(\rho(x))dx.$$

Thus it acts as $(\phi F)(x) = \phi(\rho(x))F(x)$ for each $x \in \mathbb{R}^d$.

Definition 4.3 (*The Nelson Hamiltonian*) The Nelson Hamiltonian is defined by

$$H_{\text{N}} = H_{\text{N},0} + \phi,$$

where

$$H_{\text{N},0} = H_{\text{p}} \otimes \mathbb{1} + \mathbb{1} \otimes H_{\text{f}}.$$

In what follows, we omit tensor notation $\otimes$ and we simply write as $H_{\text{p}} + H_{\text{f}} + \phi$ unless confusion may arise.

Proposition 4.4 (Selfadjointness) *H_{N} is selfadjoint and bounded from below on $D(H_{\text{N},0})$ and essentially selfadjoint on any core of $H_{\text{N},0}$.*

Proof For any $\varepsilon > 0$, it follows that $\|\phi F\| \le \varepsilon\|H_{\text{N},0}F\| + b_\varepsilon\|F\|$ for any $F \in D(H_{\text{N},0})$. Then the proposition follows from the Kato-Rellich theorem. ■

In this section we prove the existence of the ground state of H_N with cutoffs. We can prove this in a similar way to the Pauli–Fierz Hamiltonian. We shall however give an alternative proof which is due to Gérard [17, 18]. Here we do not need the binding condition (Assumption 3.10). Instead of this the compactness of the resolvent of particle Hamiltonian H_p is required.[1]

4.1.2 Existence of the Ground State for $\sigma > 0$

In this section we show the existence of the ground state of the Nelson Hamiltonian with infrared cutoff.

Let $\sigma \geq 0$ be an infrared cutoff parameter and we define the field operator with infrared cutoff σ by $\phi_\sigma = \phi(\rho_\sigma)$, where $\rho_\sigma = \rho \mathbb{1}_{\sigma \leq \omega}$. Here and in what follows we simply denote $\sigma \leq \omega$ for the subscript of $\mathbb{1}_{\sigma \leq \omega}$ instead of $\{k \in \mathbb{R}^d | \sigma \leq \omega(k)\}$. We define H_σ by H_N with ϕ replaced by ϕ_σ:

$$H_\sigma = H_{N,0} + \phi_\sigma.$$

Simply we set

$$H = H_{\sigma=0} = H_N.$$

For each $\sigma \geq 0$, H_σ is selfadjoint on $D(H_\sigma) = D(H_N)$. Let $E_\sigma = \inf \mathrm{sp}(H_\sigma)$ and $E = \inf \mathrm{sp}(H)$.

Lemma 4.5 *H_σ converges to H as $\sigma \to 0$ in the norm resolvent sense. In particular* $\lim_{\sigma \to 0} E_\sigma = E$.

Proof By the bound $\|(H_f + \mathbb{1})(H_\sigma - z)^{-1}\| < C$ with some constant C and $z \in \mathbb{C}$ with $\Im z \neq 0$, we can see that

$$\|(H_\sigma - z)^{-1} - (H - z)^{-1}\| \leq C \frac{1}{|\Im z|} (\|\mathbb{1}_{\omega < \sigma} \hat{\varphi}/\omega\| + \|\mathbb{1}_{\omega < \sigma} \hat{\varphi}/\sqrt{\omega}\|).$$

The right-hand side converges to 0 as $\sigma \to 0$, and the lemma follows. ■

Let us introduce a smooth multiplication operator $\tilde{\omega}_\sigma$ such that $\tilde{\omega}_\sigma \in C^1(\mathbb{R}^d)$, $\nabla_\mu \tilde{\omega}_\sigma \in L^\infty(\mathbb{R}^d)$ for $\mu = 1, \ldots, d$ and

$$\tilde{\omega}_\sigma(k) \begin{cases} \geq \sigma/2, & |k| < \sigma, \\ = \omega(k), & |k| \geq \sigma. \end{cases}$$

[1] In [17] more general Hamiltonian is studied;

$$K \otimes \mathbb{1} + \mathbb{1} \otimes d\Gamma(h) + \int (v(k) \otimes a^\dagger(k) + \bar{v}(k) \otimes a(k)) dk,$$

where K is a selfadjoint operator with compact resolvent, and h some positive selfadjoint operator.

We shall use technical assumptions $\tilde{\omega}_\sigma \in C^1(\mathbb{R}^d)$ and $\nabla_\mu \tilde{\omega}_\sigma \in L^\infty(\mathbb{R}^d)$ only in the proof of Lemma 4.6 below. We define Hamiltonians $\tilde{H}_\sigma$ and $\tilde{H}_{0,\sigma}$ by

$$\begin{aligned}\tilde{H}_{0,\sigma} &= H_\mathrm{p} + \mathrm{d}\Gamma(\tilde{\omega}_\sigma),\\ \tilde{H}_\sigma &= H_\mathrm{p} + \mathrm{d}\Gamma(\tilde{\omega}_\sigma) + \phi_\sigma.\end{aligned}$$

We extend the Hilbert space $\mathscr{H}$ to show the existence of a ground state of H_σ by a localization argument. Let $\mathscr{H}^{\mathrm{ex}} = \mathscr{H} \otimes \mathscr{F}$ and we define selfadjoint operators by

$$\begin{aligned}\tilde{H}_{0,\sigma}^{\mathrm{ex}} &= \tilde{H}_{0,\sigma} \otimes \mathbb{1}_{\mathscr{F}} + \mathbb{1}_{\mathscr{H}} \otimes \mathrm{d}\Gamma(\tilde{\omega}_\sigma),\\ \tilde{H}_\sigma^{\mathrm{ex}} &= \tilde{H}_\sigma \otimes \mathbb{1}_{\mathscr{F}} + \mathbb{1}_{\mathscr{H}} \otimes \mathrm{d}\Gamma(\tilde{\omega}_\sigma).\end{aligned}$$

Let $j = (j_0, j_\infty) \in l_{\mathrm{loc}}(\mathbb{R}^d)$. We define $j_{*,R}$ and $\hat{j}_{*,R}$ by $j_{*,R} = j_*(\cdot/R)$ and $\hat{j}_{*,R} = j_{*,R}(-i\nabla)$ for $* = 0, \infty$. Operator $\hat{j}_R : L^2(\mathbb{R}^d) \to L^2(\mathbb{R}^d) \oplus L^2(\mathbb{R}^d)$ is defined by $\hat{j}_R \Psi = \hat{j}_{0,R}\Psi \oplus \hat{j}_{\infty,R}\Psi$, and $\mathscr{U}_\mathrm{c} : \mathscr{F}(L^2(\mathbb{R}^d) \oplus L^2(\mathbb{R}^d)) \to \mathscr{F} \otimes \mathscr{F}$ is the unitary operator. We set $U_R = \mathscr{U}_\mathrm{c}\mathrm{d}\Gamma(\hat{j}_R) : \mathscr{F} \to \mathscr{F}(L^2(\mathbb{R}^d)) \otimes \mathscr{F}(L^2(\mathbb{R}^d))$. We have

$$U_R \left(\prod_{j=1}^n a^\dagger(f_j) \right) U_R^{-1} = \prod_{j=1}^n \left(a^\dagger(\hat{j}_{0,R} f_j) \otimes \mathbb{1} + \mathbb{1} \otimes a^\dagger(\hat{j}_{\infty,R} f_j) \right).$$

$\mathbb{1}_{L^2} \otimes U_R$ is also denoted by the same symbol U_R in what follows. We want to estimate an error term of $U_R \chi_1(\tilde{H}_\sigma) - \chi_1(\tilde{H}_\sigma^{\mathrm{ex}})U_R$.

Lemma 4.6 *Let $\chi_1, \chi_2 \in C_0^\infty(\mathbb{R})$. Then*

$$\lim_{R\to 0} \left\| \left(\chi_1(\tilde{H}_\sigma^{\mathrm{ex}})U_R - U_R\chi_1(\tilde{H}_\sigma) \right) \chi_2(\tilde{H}_\sigma) \right\| = 0.$$

I.e., $U_R\chi_1(\tilde{H}_\sigma)\chi_2(\tilde{H}_\sigma) = \chi_1(\tilde{H}_\sigma^{\mathrm{ex}})U_R\chi_2(\tilde{H}_s) + o(R^0)$.

Proof Let $\tilde{\chi}_1$ be an almost analytic extension of χ_1 such that $\tilde{\chi}_1(x) = \chi_1(x)$ for $x \in \mathbb{R}$, $\tilde{\chi}_1 \in C_0^\infty(\mathbb{C})$ and $|\partial_{\bar{z}}\tilde{\chi}_1(z)| \le C_n|\Im z|^n$ for $n \in \mathbb{N}$. We have

$$U_R\chi_1(\tilde{H}_\sigma) - \chi_1(\tilde{H}_\sigma^{\mathrm{ex}})U_R = \frac{-1}{\pi}\int_{\mathbb{C}} \partial_{\bar{z}}\tilde{\chi}_1(z)(\tilde{H}_\sigma^{\mathrm{ex}} - z)^{-1}(U_R\tilde{H}_\sigma - \tilde{H}_\sigma^{\mathrm{ex}}U_R)(\tilde{H}_\sigma - z)^{-1}dxdy. \tag{4.1}$$

Let us estimate the integrand in (4.1). We have

$$U_R\tilde{H}_\sigma - \tilde{H}_\sigma^{\mathrm{ex}}U_R = U_R\tilde{H}_{0,\sigma} - \tilde{H}_{0,\sigma}^{\mathrm{ex}}U_R - (\phi_\sigma \otimes \mathbb{1}_{\mathscr{F}})U_R + U_R\phi_\sigma. \tag{4.2}$$

Since the first term on the right-hand side of (4.2) is given by

$$U_R\tilde{H}_{0,\sigma} - \tilde{H}_{0,\sigma}^{\mathrm{ex}}U_R = -d\Gamma(\hat{j}_R, (\tilde{\omega}_\sigma \oplus \tilde{\omega}_\sigma)\hat{j}_R - \hat{j}_R\tilde{\omega}_\sigma),$$

we already shown that $\|d\Gamma(\hat{j}_R,(\tilde{\omega}_\sigma \oplus \tilde{\omega}_\sigma)\hat{j}_R - \hat{j}_R\tilde{\omega}_\sigma)(N+\mathbb{1})^{-1}\| \to 0$ as $R \to \infty$ in the proof of Lemma 3.15, where we need $\tilde{\omega}_\sigma \in C^1(\mathbb{R}^d)$ and $\nabla_\mu \tilde{\omega}_\sigma \in L^\infty(\mathbb{R}^d)$ for $\mu = 1, \ldots, d$. Let us consider the second term of (4.2). We have

$$\phi_\sigma \otimes \mathbb{1}_{\mathscr{F}} U_R - U_R \phi_\sigma = \left(\phi((1-\hat{j}_{0,R})\rho_\sigma) \otimes \mathbb{1}_{\mathscr{F}} - \mathbb{1}_{\mathscr{H}} \otimes \phi(\hat{j}_{\infty,R}\rho_\sigma)\right) U_R,$$

but it is also proven in the proof of Lemma 3.15 that as $R \to \infty$,

$$\left\| \left(\phi((1-\hat{j}_{0,R})\rho_\sigma) \otimes \mathbb{1}_{\mathscr{F}} - \mathbb{1}_{\mathscr{H}} \otimes \phi(\hat{j}_{\infty,R}\rho_\sigma)\right) U_R (N+\mathbb{1})^{-1} \right\| \to 0.$$

We have $\|(U_R \tilde{H}_\sigma - \tilde{H}_\sigma^{\mathrm{ex}} U_R)(N+\mathbb{1})^{-1}\| = o(R^0)$. Since $N \leq \frac{1}{\sigma} H_{\mathrm{f}}$, we see that

$$\|NF\| \leq C\|(\tilde{H}_\sigma + \mathbb{1})F\|$$

with some constant C. The integrand of (4.1) can be estimated as

$$\begin{aligned}
&|\partial_{\bar{z}}\tilde{\chi}_1(z)| \|(\tilde{H}_\sigma^{\mathrm{ex}} - z)^{-1}(U_R \tilde{H}_\sigma - \tilde{H}_\sigma^{\mathrm{ex}} U_R)(\tilde{H}_\sigma - z)^{-1}\chi_2(\tilde{H}_\sigma)\| \\
&\leq \frac{|\partial_{\bar{z}}\tilde{\chi}_1(z)|}{|\Im z|} \|(U_R \tilde{H}_\sigma - \tilde{H}_\sigma^{\mathrm{ex}} U_R)(N+\mathbb{1})^{-1}\| \|(N+\mathbb{1})(\tilde{H}_\sigma - z)^{-1}\| \|\chi_2(\tilde{H}_\sigma)\| \\
&\leq C\|(U_R \tilde{H}_\sigma - \tilde{H}_\sigma^{\mathrm{ex}} U_R)(N+\mathbb{1})^{-1}\| \frac{|\partial_{\bar{z}}\tilde{\chi}_1(z)|}{|\Im z|} \left(1 + \frac{1}{|\Im z|}\right),
\end{aligned}$$

where C is a constant independent of z and R. Hence the lemma follows. ∎

Lemma 4.7 *Let* $\tilde{E}_\sigma = \inf \mathrm{sp}(\tilde{H}_\sigma)$ *and* $\chi \in C_0^\infty(\mathbb{R})$ *with* $\mathrm{supp}\,\chi \subset (-\infty, \tilde{E}_\sigma + \sigma/2)$ *and* $\chi \geq 0$. *Then* $\chi(\tilde{H}_\sigma)$ *is compact. In particular,* $\tilde{H}_\sigma$ *has a ground state.*

Proof Let P_Ω be the projection to $\{\alpha\Omega | \alpha \in \mathbb{C}\}$. Since $\mathrm{supp}\chi \subset (-\infty, \tilde{E}_\sigma + \sigma/2)$ and $\mathbb{1}_{\mathscr{H}} \otimes \mathrm{d}\Gamma(\tilde{\omega}_\sigma)\lceil_{\oplus_{n\geq 1}\mathscr{F}^{(n)}} \geq \sigma$, we see that

$$\chi(\tilde{H}_\sigma^{\mathrm{ex}}) = (\mathbb{1}_{\mathscr{H}} \otimes P_\Omega)\chi(\tilde{H}_\sigma^{\mathrm{ex}}). \tag{4.3}$$

We also see that $U_R^*(\mathbb{1}_{\mathscr{H}} \otimes P_\Omega)U_R = \mathbb{1}_{L^2} \otimes \Gamma(\hat{j}_{0,R}^2)$. We set $\mathbb{1}_{L^2} \otimes \Gamma(\hat{j}_{0,R}^2) = \Gamma(\hat{j}_{0,R}^2)$ for simplicity. From Lemma 4.6 and (4.3) it follows that

$$\begin{aligned}
\chi(\tilde{H}_\sigma) &= U_R^* U_R \chi^{1/2}(\tilde{H}_\sigma)\chi^{1/2}(\tilde{H}_\sigma) \\
&= U_R^*(\mathbb{1}_{\mathscr{H}} \otimes P_\Omega)\chi^{1/2}(\tilde{H}_\sigma^{\mathrm{ex}})U_R\chi^{1/2}(\tilde{H}_\sigma) + o(R^0).
\end{aligned}$$

By Lemma 4.6 again we have

$$\chi(\tilde{H}_\sigma) = U_R^*(\mathbb{1}_{\mathscr{H}} \otimes P_\Omega)U_R\chi(\tilde{H}_\sigma) + o(R^0) = \Gamma(\hat{j}_{0,R}^2)\chi(\tilde{H}_\sigma) + o(R^0). \tag{4.4}$$

We shall show that $\Gamma(\hat{j}_{0,R}^2)\chi(\tilde{H}_\sigma)$ is compact. Since

$$\lim_{n\to\infty}\left\|\Gamma(\hat{j}_{0,R}^2)\chi(\tilde{H}_\sigma)-\sum_{k=0}^{n}\mathbb{1}_{\{k\}}(N)\Gamma(\hat{j}_{0,R}^2)\chi(\tilde{H}_\sigma)\right\|\le\lim_{n\to\infty}\frac{\|\Gamma(\hat{j}_{0,R}^2)N\chi(\tilde{H}_\sigma)\|}{n+1}=0,$$

it suffices to show that $\mathbb{1}_{\{k\}}(N)\Gamma(\hat{j}_{0,R}^2)\chi(\tilde{H}_\sigma)$ is a compact operator for each k. Since $\mathbb{1}_{\{k\}}(N)\Gamma(\hat{j}_{0,R}^2)\chi(\tilde{H}_\sigma)=KL$ with

$$K=(H_{\mathrm{p}}+\mathbb{1})^{-1/2}\otimes\Gamma(\hat{j}_{0,R}^2)(\mathrm{d}\Gamma(\tilde{\omega}_\sigma)+\mathbb{1})^{-1/2}\mathbb{1}_{\{k\}}(N),$$
$$L=\left((H_{\mathrm{p}}+\mathbb{1})^{1/2}\otimes(\mathrm{d}\Gamma(\tilde{\omega}_\sigma)+\mathbb{1})^{1/2}\right)\chi(\tilde{H}_\sigma).$$

Here K is compact and L bounded. Thus $\mathbb{1}_{\{k\}}(N)\Gamma(\hat{j}_{0,R}^2)\chi(\tilde{H}_\sigma)$ is compact and $\chi(\tilde{H}_\sigma)$ is also compact by (4.4). ■

We can show the existence of the ground state for $\sigma>0$.

Theorem 4.8 (Gérard [17]) *Let $\sigma>0$. Then H_σ has a ground state.*

Proof Let $U_\sigma:\mathscr{F}(L^2(\mathbb{R}^d))\to\mathscr{F}(L^2(\{\sigma\le\omega\}))\otimes\mathscr{F}(L^2(\sigma>\omega))\mathscr{F}_1\otimes\mathscr{F}_2$ be the unitary operator. Since

$$(\mathbb{1}_{L^2}\otimes U_\sigma)H_\sigma(\mathbb{1}_{L^2}\otimes U_\sigma^*)=\mathbb{1}_{L^2\otimes\mathscr{F}_1}\otimes\mathrm{d}\Gamma(\omega\lceil_{\sigma>\omega})+K_\sigma\otimes\mathbb{1}_{\mathscr{F}_2},\tag{4.5}$$

where $K_\sigma=H_{\mathrm{p}}+\mathrm{d}\Gamma(\omega\lceil_{\sigma\le\omega})+\phi_\sigma$. We know that $\mathrm{d}\Gamma(\omega\lceil_{\sigma>\omega})$ has a ground state. It is sufficient to show that K_σ also has a ground state. It can be seen that

$$(\mathbb{1}_{L^2}\otimes U_\sigma)\tilde{H}_\sigma(\mathbb{1}_{L^2}\otimes U_\sigma^*)=\mathbb{1}_{L^2\otimes\mathscr{F}_1}\otimes\mathrm{d}\Gamma(\tilde{\omega}_\sigma)+K_\sigma\otimes\mathbb{1}_{\mathscr{F}_2}.\tag{4.6}$$

Since each $\tilde{H}_\sigma$ and $\mathrm{d}\Gamma(\tilde{\omega}_\sigma)$ has a ground state, K_σ also has a ground state. The proof is complete. ■

4.1.3 Existence of the Ground State for $\sigma=0$

In this section we show the existence of the ground state of the Nelson Hamiltonian without infrared cutoff, i.e., we show that H_σ has a ground state for $\sigma=0$.

Let Ψ_σ be a normalized ground state of H_σ for $\sigma>0$. In the same way as that of the Pauli–Fierz Hamiltonian we can derive

$$a(f)\Psi_\sigma=\int_{\mathbb{R}^d}f(k)(H_\sigma-E_\sigma+\omega(k))^{-1}\Psi_\sigma dk\tag{4.7}$$

by using asymptotic fields. For each $k\in\mathbb{R}^d$ we define the notation $(a\Phi)^{(n)}(k)$ by

$$(a\Phi)^{(n)}(k) = \sqrt{n}\Phi^{(n)}(k, \cdot)$$

and

$$(a\Phi)(k) = \oplus_{n=0}^{\infty}\sqrt{n}\Phi^{(n)}(k, \cdot). \tag{4.8}$$

Remark 4.9 Map $\mathbb{R}^d \ni k \mapsto \Psi^{(n)}(k, \ldots)$ can be regarded as an $\mathscr{H}^{(n-1)}$-valued L^2-function. In general $\Psi^{(n)}(k, \ldots)$ is not defined pointwise, because $\Psi^{(n)}$ is a function in $\mathscr{H}^{(n)} = L^2(\mathbb{R}_x^d \times \mathbb{R}_k^{dn})$.

$$\Psi^{(n)} = \Phi^{(n)} \quad \text{if and only if} \quad \|\Psi^{(n)}(k, \ldots) - \Phi^{(n)}(k, \ldots)\|_{\mathscr{H}^{(n-1)}} = 0, \quad a.e.k.$$

By this observation we see that if $\Psi^{(n)} = \Phi^{(n)}$ as an $\mathscr{H}^{(n-1)}$-valued L^2-function, $(a\Psi)^{(n)}(k) = (a\Phi)^{(n)}(k)$ in $\mathscr{H}^{(n-1)}$ for $k \in \mathbb{R}^d \setminus N$ with some null set N depending on Ψ and Φ.

Hence

$$a(f)\Psi_\sigma = \int_{\mathbb{R}^d} f(k)(a\Psi_\sigma)(k, \cdot)dk \tag{4.9}$$

follows from the definition of $a(f)$. Here $f \in L^2(\mathbb{R}^d)$ is arbitrary. Comparing (4.7) with (4.9) we obtain that

$$(a\Psi_\sigma)(k, \cdot) = \frac{1}{\sqrt{2}}(H_\sigma - E_\sigma + \omega(k))^{-1}\rho_\sigma(k)\Psi_\sigma(\cdot) \quad a.e.\ k \in \mathbb{R}^d.$$

Let $h : \mathbb{R}^d \to \mathbb{C}$ be measurable and non-negative. For every $\Psi, \Phi \in D(\mathrm{d}\Gamma(h)^{1/2})$ we have

$$\begin{aligned}
(\Psi^{(n)}, (\mathrm{d}\Gamma(h)\Phi)^{(n)}) &= \sum_{i=1}^{n} \int_{\mathbb{R}^{nd}} \Psi^{(n)}(k_1, \ldots, k_n)\Phi^{(n)}(k_1, \ldots, k_n)h(k_i)dk_1 \cdots dk_n \\
&= n \int_{\mathbb{R}^{nd}} \Psi^{(n)}(k_1, \ldots, k_n)\Phi^{(n)}(k_1, \ldots, k_n)h(k_1)dk_1 \cdots dk_n \\
&= \int_{\mathbb{R}^d} ((a\Psi)^{(n)}(k, \cdot), h(k)(a\Phi)^{(n)}(k, \cdot))_{\mathscr{F}^{(n-1)}}dk.
\end{aligned}$$

Here we used the symmetry of $\Psi(k_1, \ldots, k_n)$ and $\Phi(k_1, \ldots, k_n)$ with respect to permutations of $k_1, \ldots, k_n$. Thus in general we have

$$(\Psi, \mathrm{d}\Gamma(h)\Phi)_{\mathscr{F}} = \int_{\mathbb{R}^d} h(k)\|(a\Psi)(k)\|_{\mathscr{F}}^2 dk.$$

In particular we have the pull-through formula:

$$\|N^{1/2}\Psi_\sigma\|^2 = \frac{1}{2}\int_{\mathbb{R}^d} \|(H_\sigma - E_\sigma + \omega(k))^{-1}\rho_\sigma(k)\Psi_\sigma\|^2 dk.$$

Definition 4.10 (*Infrared regular and singular condition*)

$$\int_{\mathbb{R}^d} \frac{|\hat{\varphi}(k)|^2}{\omega(k)^3} dk < \infty \tag{4.10}$$

is called infrared regular condition and

$$\int_{\mathbb{R}^d} \frac{|\hat{\varphi}(k)|^2}{\omega(k)^3} dk = \infty \tag{4.11}$$

is called infrared singular condition.

We set the integral on the left-hand side of (4.10) by I.

Lemma 4.11 *Suppose $I < \infty$. Then $\Psi_\sigma \in D(N^{1/2})$ for each $\sigma > 0$ and it follows that* $\sup_{0<\sigma\le 1}\|N^{1/2}\Psi_\sigma\| < \infty$.

Proof We have $\|N^{1/2}\Psi_\sigma\|^2 \le I/2 < \infty$. Thus the lemma follows. ■

We have already seen that $\sigma \mapsto E_\sigma$ is continuous which can be derived from the fact that the resolvent $(H_\sigma - z)^{-1}$ is continuous in σ in the operator norm. We can prove a stronger statement.

Lemma 4.12 *Suppose $I < \infty$. Then $|E_\sigma - E| \in o(\sigma^1)$, i.e, $\lim_{\sigma\to 0}|E_\sigma - E|/\sigma = 0$.*

Proof Let $0 < \sigma < \tau < 1$. We have

$$E_\tau - E_\sigma \le (\Psi_\sigma, H_\tau\Psi_\sigma) - (\Psi_\sigma, H_\sigma\Psi_\sigma) = (\Psi_\sigma, \phi(\rho_\tau - \rho_\sigma)\Psi_\sigma),$$
$$E_\sigma - E_\tau \le (\Psi_\tau, H_\sigma\Psi_\tau) - (\Psi_\tau, H_\tau\Psi_\tau) = (\Psi_\tau, \phi(\rho_\sigma - \rho_\tau)\Psi_\tau).$$

Since $\|\phi(\rho_\sigma - \rho_\tau)\Phi\| \le \|\rho_\sigma - \rho_\tau\|\|(N + \mathbb{1})^{1/2}\Phi\|/\sqrt{2}$ and

$$\|\rho_\sigma - \rho_\tau\|^2 = \int_{\tau\le\omega<\sigma} \frac{|\hat{\varphi}(k)|^2}{\omega(k)} dk,$$

we have

$$|E_\tau - E_\sigma|^2 \le C\int_{\sigma>\omega} \frac{|\hat{\varphi}(k)|^2}{\omega(k)} dk.$$

Noticing that $\lim_{\tau\to 0} E_\tau = E$, we take the limit and we obtain that

$$\frac{|E - E_\sigma|^2}{\sigma^2} \le C\int_{\sigma>\omega} \frac{|\hat{\varphi}(k)|^2}{\omega(k)^3} dk.$$

From this $|E - E_\sigma| \in o(\sigma^1)$ follows. ■

Lemma 4.13 *Suppose $I < \infty$. Then*

$$\lim_{\sigma\to 0}\int_{\mathbb{R}^d}\left\|(a\Psi_\sigma)(k)-\frac{1}{\sqrt{2}}(H-E+\omega(k))^{-1}\rho(k)\Psi_\sigma\right\|^2 dk=0.$$

Proof By the pull-through formula, we have

$$(a\Psi_\sigma)(k)-\frac{1}{\sqrt{2}}(H-E+\omega(k))^{-1}\rho(k)\Psi_\sigma=\frac{1}{\sqrt{2}}(R_1(k)+R_2(k)+R_3(k)),$$

where

$$\begin{aligned}
R_1(k)&=(H_\sigma-E_\sigma+\omega(k))^{-1}\phi(\rho-\rho_\sigma)(H_\sigma-E_\sigma+\omega(k))^{-1}\rho_\sigma(k)\Psi_\sigma,\\
R_2(k)&=(H_\sigma-E_\sigma+\omega(k))^{-1}(E-E_\sigma)(H_\sigma-E_\sigma+\omega(k))^{-1}\rho_\sigma(k)\Psi_\sigma,\\
R_3(k)&=(H-E+\omega(k))^{-1}(\rho_\sigma(k)-\rho(k))\Psi_\sigma.
\end{aligned}$$

Firstly let us consider R_1. We have

$$\|R_1(k)\|\le\frac{1}{\omega(k)}\|\rho-\rho_\sigma\|\|(N+\mathbb{1})^{1/2}(H_\sigma-E_\sigma+\omega(k))^{-1}\rho_\sigma(k)\Psi_\sigma\|$$

and we shall show that

$$\|(N+\mathbb{1})^{1/2}(H_\sigma-E_\sigma+\omega(k))^{-1}\rho_\sigma(k)\Psi_\sigma\|\le\max\left\{\frac{1}{\sqrt{\omega(k)}},\frac{1}{\omega(k)}\right\}\frac{|\hat{\varphi}(k)|}{\sqrt{\omega(k)}} \tag{4.12}$$

in Lemma 4.14 below. Hence

$$\int_{\mathbb{R}^d}\|R_1(k)\|^2dk\le\int_{\sigma\le\omega}\left(\frac{1}{\omega(k)}\left\|\mathbb{1}_{\sigma>\omega}\frac{\hat{\varphi}}{\sqrt{\omega}}\right\|\max\left\{\frac{1}{\sqrt{\omega(k)}},\frac{1}{\omega(k)}\right\}\frac{|\hat{\varphi}(k)|}{\sqrt{\omega(k)}}\right)^2dk.$$

It is trivial to see that

$$\int_{1\le\omega}\|R_1(k)\|^2dk\le\int_{1\le\omega}\frac{|\hat{\varphi}(k)|^2}{\omega(k)^4}dk\int_{\sigma>\omega}\frac{|\hat{\varphi}(k)|^2}{\omega(k)}dk\to 0$$

as $\sigma\to 0$. Using $1/\sigma>1/\omega$ on$\{\sigma<\omega\}$ and $1/\sigma\le 1/\omega$ on$\{\sigma\ge\omega\}$, we can see that

$$\int_{\sigma\le\omega<1}\|R_1(k)\|^2dk \le \int_{\sigma\le\omega<1}\frac{1}{\omega(k)^2}\left\|\mathbb{1}_{\sigma>\omega}\frac{\hat{\varphi}}{\sqrt{\omega}}\right\|^2\frac{|\hat{\varphi}(k)|^2}{\omega(k)^3}dk$$
$$\le \int_{\sigma\le\omega<1}\left\|\mathbb{1}_{\sigma>\omega}\frac{\hat{\varphi}}{\sigma\sqrt{\omega}}\right\|^2\frac{|\hat{\varphi}(k)|^2}{\omega(k)^3}dk \le \int_{\sigma\le\omega<1}\frac{|\hat{\varphi}(k)|^2}{\omega(k)^3}dk\int_{\sigma>\omega}\frac{|\hat{\varphi}(k)|^2}{\omega(k)^3}dk \to 0$$

as $\sigma \to 0$. Hence $\lim_{\sigma\to 0}\int_{\mathbb{R}^d}\|R_1(k)\|^2dk = 0$ follows. Next we consider R_2. We have

$$\int_{\mathbb{R}^d}|R_2(k)|^2dk \le \int_{\sigma\le\omega}\frac{|E-E_\sigma|^2}{\omega(k)^4}\frac{|\hat{\varphi}(k)|^2}{\omega(k)}dk \le \frac{|E-E_\sigma|^2}{\sigma^2}\int_{\sigma\le\omega}\frac{|\hat{\varphi}(k)|^2}{\omega(k)^3}dk \to 0$$

as $\sigma \to 0$. Finally we have

$$\int_{\mathbb{R}^d}\|R_3(k)\|^2dk \le \int_{\sigma>\omega}\frac{|\hat{\varphi}(k)|^2}{\omega(k)^3}dk \to 0$$

as $\sigma \to 0$. Hence the lemma follows. ■

Lemma 4.14 *Inequality (4.12) holds.*

Proof Let $f \in C_0^\infty(\mathbb{R}^d)$ and fix $k \in \mathbb{R}^d$. Set $R_\rho = (H_\sigma - E_\sigma + \omega(k))^{-1}\rho_\sigma(k)$ and $R = (H_\sigma - E_\sigma + \omega(k))^{-1}$. We have

$$a(f)R_\rho\Psi_\sigma = Ra(f)\Psi_\sigma + R(-a(\omega f) - (\bar{f}, \rho_\sigma/\sqrt{2}))R_\rho\Psi_\sigma.$$

From this it follows that

$$\left((H_\sigma - E_\sigma + \omega(k))^{-1}a(f) + a(\omega f)\right)R_\rho\Psi_\sigma = \left(a(f) - (\bar{f}, \rho_\sigma/\sqrt{2})\right)R_\rho\Psi_\sigma.$$

Let $\Phi \in \mathscr{F}_{\rm fin} \cap D(H_{\rm f})$. We have

$$\left(\Phi, ((H_\sigma - E_\sigma + \omega(k))^{-1}a(f) + a(\omega f))R_\rho\Psi_\sigma\right) = (\Phi, (a(f) - (\bar{f}, \rho_\sigma/\sqrt{2}))R_\rho\Psi_\sigma).$$

By the definition: $(a(f)\Psi)(k') = \int_{\mathbb{R}^d} f(k)(a\Psi)(k,k')dk$, we have

$$\int_{\mathbb{R}^d} f(k')(\Phi, \left(H_\sigma - E_\sigma + \omega(k) + \omega(k')\right)a(R_\rho\Psi_\sigma)(k'))dk'$$
$$= \int_{\mathbb{R}^d} f(k')(\Phi, a(R_\rho\Psi_\sigma)(k') - \frac{\rho_\sigma(k')}{\sqrt{2}}R_\rho\Psi_\sigma)dk'.$$

Hence we have the identity:

$$\left(H_\sigma - E_\sigma + \omega(k) + \omega(k')\right)a(R_\rho\Psi_\sigma)(k') = a(R_\rho\Psi_\sigma)(k') - \frac{\rho_\sigma(k')}{\sqrt{2}}R_\rho\Psi_\sigma$$

a.e. $k' \in \mathbb{R}^d$, and

$$a(R_\rho \Psi_\sigma)(k') = \left(H_\sigma - E_\sigma + \omega(k) + \omega(k')\right)^{-1} \left(a(R_\rho \Psi_\sigma)(k') - \frac{\rho_\sigma(k')}{\sqrt{2}} R_\rho \Psi_\sigma \right).$$

Using this we have

$$\begin{aligned}
\|N^{1/2} R_\rho \Psi_\sigma\|^2 &= \int_{\mathbb{R}^d} \|a(R_\rho \Psi_\sigma)(k')\|^2 dk' \\
&= \int_{\mathbb{R}^d} \left\| \left(H_\sigma - E_\sigma + \omega(k) + \omega(k')\right)^{-1} \left(a(R_\rho \Psi_\sigma)(k') - \frac{\rho_\sigma(k')}{\sqrt{2}} R_\rho \Psi_\sigma \right) \right\|^2 dk' \\
&\leq a + b,
\end{aligned}$$

where

$$\begin{aligned}
a &= 2 \int_{\mathbb{R}^d} \left\| \left(H_\sigma - E_\sigma + \omega(k) + \omega(k')\right)^{-1} a(R_\rho \Psi_\sigma)(k') \right\|^2 dk', \\
b &= 2 \int_{\mathbb{R}^d} \left\| \left(H_\sigma - E_\sigma + \omega(k) + \omega(k')\right)^{-1} \frac{\rho_\sigma(k')}{\sqrt{2}} R_\rho \Psi_\sigma \right\|^2 dk'.
\end{aligned}$$

We have

$$a \leq \frac{2}{\omega(k)^2} \|N^{1/2} \Psi_\sigma\|^2 \leq \frac{2}{\omega(k)^2} \int_{\mathbb{R}^d} \frac{|\hat{\varphi}(k')|^2}{\omega(k')^3} dk' \|\Psi_\sigma\|^2, \quad b \leq \frac{|\hat{\varphi}(k)|^2}{\omega(k)^3} \int_{\mathbb{R}^d} \frac{|\hat{\varphi}(k')|^2}{\omega(k')^3} dk' \|\Psi_\sigma\|^2.$$

The proof is complete. ■

Let $F \in C_0^\infty(\mathbb{R}^d)$ be such that $0 \leq F(k) \leq 1$ for all $k \in \mathbb{R}^d$ and $F(0) = 1$. We set $\hat{F}_R = F(-i\nabla_k/R)$. Hence $\hat{F}_R$ acts as $\hat{F}_R f(k) = (2\pi)^{-d/2} \int_{\mathbb{R}^d} \check{F}(s) f(k + s/R) ds$ and the second quantization of $\hat{F}_R$ satisfies that

$$(\Phi, \mathrm{d}\Gamma(\hat{F}_R)\Psi)_{\mathscr{F}} = \int_{\mathbb{R}^d} ((a\Phi)(k), \hat{F}_R(a\Psi)(k))_{\mathscr{F}} dk$$

and $\hat{F}_R(a\Psi)(k) = (2\pi)^{-d/2} \int_{\mathbb{R}^d} \check{F}(s)(a\Psi)(k + s/R) ds$.

Lemma 4.15 *Suppose $I < \infty$. Then*

$$\|\mathrm{d}\Gamma(\mathbb{1} - \hat{F}_R)^{1/2} \Psi_\sigma\| = o(R^0) + o(\sigma^0).$$

Proof Since $\Psi_\sigma \in D(N^{1/2})$, we see $\Psi_\sigma \in D(\mathrm{d}\Gamma(\mathbb{1} - \hat{F}_R)^{1/2})$ and

$$\|\mathrm{d}\Gamma(\mathbb{1} - \hat{F}_R)^{1/2} \Psi_\sigma\|^2 = \int_{\mathbb{R}^d} ((a\Psi_\sigma)(k), (\mathbb{1} - \hat{F}_R)(a\Psi_\sigma)(k)) dk. \tag{4.13}$$

By the Schwarz inequality we have

$$(4.13) \leq \left(\int_{\mathbb{R}^d} \|(a\Psi_\sigma)(k)\|^2 dk\right)^{1/2} \left(\int_{\mathbb{R}^d} \|(\mathbb{1} - \hat{F}_R)(a\Psi_\sigma)(k)\|^2 dk\right)^{1/2}$$
$$= \|N^{1/2}\Psi_\sigma\| \left(\int_{\mathbb{R}^d} \|(\mathbb{1} - \hat{F}_R)(H - E + \omega(k))^{-1}\rho(k)\Psi_\sigma\|^2 dk\right)^{1/2} + o(\sigma^0).$$

Here we used Lemma 4.13. We see that

$$\|d\Gamma(\mathbb{1} - \hat{F}_R)^{1/2}\Psi_\sigma\|^2$$
$$\leq \|N^{1/2}\Psi_\sigma\| \|\Psi_\sigma\| \int_{\mathbb{R}^d} \|(\mathbb{1} - \hat{F}_R)(H - E + \omega(k))^{-1}\rho(k)\|^2 dk + o(\sigma^0).$$

Now we shall show that $(\mathbb{1} - \hat{F}_R)(H - E + \omega(k))^{-1}\rho(k) \in L^2(\mathbb{R}^d_k, B(\mathscr{H}))$ satisfies

$$\int_{\mathbb{R}^d} \|(\mathbb{1} - \hat{F}_R)(H - E + \omega(k))^{-1}\rho(k)\|^2 dk = o(R^0).$$

Since $\int_{\mathbb{R}^d} \check{F}(s)ds = (2\pi)^{d/2}$, we note that

$$(\mathbb{1} - \hat{F}_R)(H - E + \omega(k))^{-1}\rho(k)$$
$$= (2\pi)^{-d/2} \int_{\mathbb{R}^d} \check{F}(s) \left((H - E + \omega(k))^{-1}\rho(k) - (H - E + \omega(k + \frac{s}{R}))^{-1}\rho(k + \frac{s}{R})\right) ds.$$

We have

$$\int_{\mathbb{R}^d} \|(\mathbb{1} - \hat{F}_R)(H - E + \omega(k))^{-1}\rho(k)\|^2 dk \leq (2\pi)^{-d} \int_{\mathbb{R}^d} |\check{F}(s)|^2 A(\frac{s}{R}) ds,$$

where

$$A(\frac{s}{R}) = \int_{\mathbb{R}^d} \left\|(H - E + \omega(k))^{-1}\rho(k) - (H - E + \omega(k + \frac{s}{R}))^{-1}\rho(k + \frac{s}{R})\right\|^2 dk.$$

Since $A(s/R) \leq \int_{\mathbb{R}^d} |\hat{\varphi}(k)|^2/\omega(k)^3 dk < \infty$ and $\int_{\mathbb{R}^d} |\hat{F}(s)|^2 ds < \infty$, it is sufficient to show that $\lim_{R\to\infty} A(s/R) = 0$ for each s by the Lebesgue dominated convergence theorem. Set $m = s/R$. We divide $A(m)$ into three regions as $A(m) = \int_{\mathbb{R}^d} \ldots dk = \int_{|k|\leq\alpha} \ldots dk + \int_{\alpha<|k|<\beta} \ldots dk + \int_{\beta\geq|k|} \ldots dk$ for $0 < \alpha < \beta$. Fix α and β, and assume that $|m| < \alpha$. We have

$$\int_{|k|\leq\alpha} \|(H - E + \omega(k))^{-1}\rho(k) - (H - E + \omega(k + m))^{-1}\rho(k + m)\|^2 dk$$
$$\leq 2\int_{|k|\leq\alpha} \frac{|\hat{\varphi}(k)|^2}{\omega(k)^3} dk + 2\int_{|k|\leq\alpha+|m|} \frac{|\hat{\varphi}(k)|^2}{\omega(k)^3} dk.$$

In the same way we have

$$\int_{\beta\le|k|}\|(H-E+\omega(k))^{-1}\rho(k)-(H-E+\omega(k+m))^{-1}\rho(k+m)\|^2dk$$
$$\le 2\int_{\beta\le|k|}\frac{|\hat{\varphi}(k)|^2}{\omega(k)^3}dk+2\int_{\beta-|m|\le|k|}\frac{|\hat{\varphi}(k)|^2}{\omega(k)^3}dk.$$

Next we estimate the integral $\int_{\alpha<|k|<\beta}\dots dk$. Since

$$\begin{aligned}&(H-E+\omega(k))^{-1}\rho(k)-(H-E+\omega(k+m))^{-1}\rho(k+m)\\&=(H-E+\omega(k))^{-1}\rho(k)(\omega(k+m)-\omega(k))(H-E+\omega(k+m))^{-1}\rho(k)\\&+(H-E+\omega(k+m))^{-1}(\rho(k)-\rho(k+m)),\end{aligned}$$

using $|\omega(k+m)-\omega(k)|\le|m|$, we have

$$\int_{\alpha<|k|<\beta}\|(H-E+\omega(k))^{-1}\rho(k)-(H-E+\omega(k+m))^{-1}\rho(k+m)\|^2dk$$
$$\le|m|^2\int_{\alpha<|k|<\beta}\frac{|\rho(k)|^2}{\omega(k+m)^2\omega(k)^2}dk+\int_{\alpha<|k|<\beta}\frac{|\rho(k)-\rho(k+m)|^2}{\omega(k+m)^2}dk.$$

Hence we see that

$$\le\frac{|m|}{\alpha^2}\int_{\alpha<|k|<\beta}\frac{|\rho(k)|^2}{\omega(k+m)^2}dk+\frac{1}{|\alpha-|m||^2}\int_{\alpha-|m|<|k|<\beta+|m|}|\rho(k)-\rho(k+m)|^2dk$$
$$\le\frac{|m|}{\alpha^2}\int_{\alpha<|k|<\beta}\frac{|\rho(k)|^2}{\omega(k+m)^2}dk+\frac{1}{|\alpha-|m||^2}\int_{\mathbb{R}^d}|\rho(k)-\rho(k+m)|^2dk.$$

Here it follows that

$$\lim_{|m|\to 0}\int_{\alpha<|k|<\beta}\frac{|\rho(k)|^2}{\omega(k+m)^2}dk=\int_{\alpha<|k|<\beta}\frac{|\hat{\varphi}(k)|^2}{\omega(k)^3}dk<\infty,$$
$$\lim_{|m|\to 0}\int_{\mathbb{R}^d}|\rho(k)-\rho(k+m)|^2dk=0.$$

The second line can be derived from the fact that the shift in L^2 is continuous. It follows

$$\lim_{|m|\to 0}\int_{\alpha<|k|<\beta}\|(H-E+\omega(k))^{-1}\rho(k)-(H-E+\omega(k+m))^{-1}\rho(k+m)\|^2dk=0.$$

Together with them we have

$$\lim_{|m|\to 0}\int_{\mathbb{R}^d}\|(H-E+\omega(k))^{-1}\rho(k)-(H-E+\omega(k+m))^{-1}\rho(k+m)\|^2dk$$
$$\leq 4\int_{|k|\leq\alpha}\frac{|\hat{\varphi}(k)|^2}{\omega(k)^3}dk+4\int_{\beta\leq|k|}\frac{|\hat{\varphi}(k)|^2}{\omega(k)^3}dk.$$

Take sufficiently small $0<\alpha$ and sufficiently large β. The lemma is proven. ■

Theorem 4.16 (Gérard [17, Theorem 1]) *Let $I<\infty$. Then H has the ground state.*

Proof Since $\|\Psi_\sigma\|=1$ for $\sigma>0$, we can chose a sequence $\{\sigma_n\}_{n=1}^\infty$ such that $\sigma_n\to 0$ as $n\to\infty$ and $\{\Psi_{\sigma_n}\}_{n=0}^\infty$ weakly converges to some vector Φ in $\mathscr{H}$. Since H_{σ_n} converges to H in the uniform resolvent sense, we have $H\Phi=E\Phi$. Φ is a ground state of H if and only if $\Phi\neq 0$. Now we shall prove that $\Phi\neq 0$ by a contradiction. Hence we suppose $\Phi=0$. Let $F\in C_0^\infty(\mathbb{R}^d)$ be such that $0\leq F\leq 1$ and $F(0)=1$. We see that

$$\begin{aligned}\|\Psi_{\sigma_n}\|&\leq\|\Gamma(\hat{F}_R)\Psi_{\sigma_n}\|+\|(\mathbb{1}-\Gamma(\hat{F}_R))\Psi_{\sigma_n}\|\\&\leq\|\Gamma(\hat{F}_R)\mathbb{1}_{[0,\lambda]}(N)\mathbb{1}_{[0,\lambda]}(H_{\mathrm{N},0})\Psi_{\sigma_n}\|+\|\mathbb{1}_{(\lambda,\infty)}(H_{\mathrm{N},0})\Psi_{\sigma_n}\|+\|\mathbb{1}_{(\lambda,\infty)}(N)\Psi_{\sigma_n}\|\\&\quad+\|\mathbb{1}_{(\lambda,\infty)}(N)\mathbb{1}_{(\lambda,\infty)}(H_{\mathrm{N},0})\Psi_{\sigma_n}\|+\|\mathrm{d}\Gamma(\mathbb{1}-\hat{F}_R)^{1/2}\Psi_{\sigma_n}\|.\end{aligned}$$

Here we used inequality $\|(\mathbb{1}-\Gamma(\hat{F}_R))\Psi\|\leq\|\mathrm{d}\Gamma(\mathbb{1}-\hat{F}_R)^{1/2}\Psi\|$. Since we prove that $\sup_{n\in\mathbb{N}}(\Psi_{\sigma_n},N\Psi_{\sigma_n})<\infty$ by pull-through formula, and there exists a constant C such that

$$\sup_{n\in\mathbb{N}}(\Psi_{\sigma_n},H_{\mathrm{N},0}\Psi_{\sigma_n})\leq\sup_{n\in\mathbb{N}}(\Psi_{\sigma_n},(H_{\sigma_n}+C)\Psi_{\sigma_n})=\sup_{n\in\mathbb{N}}E_{\sigma_n}+C<\infty,$$

we have $\sup_{n\in\mathbb{N}}\|\mathbb{1}_{(\lambda,\infty)}(H_{\mathrm{N},0})\Psi_{\sigma_n}\|\leq c/\lambda$ and $\sup_{n\in\mathbb{N}}\|\mathbb{1}_{(\lambda,\infty)}(N)\Psi_{\sigma_n}\|\leq c/\lambda$ with some constant c. Then

$$\|\Psi_{\sigma_n}\|\leq\|\mathbb{1}_{[0,\lambda]}(N)\mathbb{1}_{[0,\lambda]}(H_{\mathrm{N},0})\Gamma(\hat{F}_R)\Psi_{\sigma_n}\|+o(R^0)+o(\sigma^0)+\frac{3c}{\lambda}$$

by Lemma 4.15. Since $(H_{\mathrm{p}}+\mathbb{1})^{-1}\otimes\Gamma(\hat{F}_R)(H_{\mathrm{f}}+\mathbb{1})^{-1}\mathbb{1}_{[0,\lambda]}(N)$ is compact by Proposition 2.30, $\Gamma(\hat{F}_R)\mathbb{1}_{[0,\lambda]}(N)\mathbb{1}_{[0,\lambda]}(H_{\mathrm{N},0})$ is also compact because of

$$\begin{aligned}&\Gamma(\hat{F}_R)\mathbb{1}_{[0,\lambda]}(N)\mathbb{1}_{[0,\lambda]}(H_{\mathrm{N},0})\\&=\left((H_{\mathrm{p}}+\mathbb{1})^{-1}\otimes\Gamma(\hat{F}_R)(H_{\mathrm{f}}+\mathbb{1})^{-1}\right)\mathbb{1}_{[0,\lambda]}(N)\left((H_{\mathrm{p}}+\mathbb{1})\otimes(H_{\mathrm{f}}+\mathbb{1})\right)\mathbb{1}_{[0,\lambda]}(H_{\mathrm{N},0})\end{aligned}$$

and $\left((H_{\mathrm{p}}+\mathbb{1})\otimes(H_{\mathrm{f}}+\mathbb{1})\right)\mathbb{1}_{[0,\lambda]}(H_{\mathrm{N},0})$ is bounded. Thus

$$\lim_{n\to\infty}\|\Gamma(\hat{F}_R)\mathbb{1}_{[0,\lambda]}(N)\mathbb{1}_{[0,\lambda]}(H_{\mathrm{N},0})\Psi_{\sigma_n}\|=0.$$

Hence we have $1 = \lim_{n\to\infty} \|\Psi_{\sigma_n}\| \le o(R^0) + 3c/\lambda < 1$ for sufficiently large R and λ. Thus we have the contradiction. Then $\Phi \neq 0$ and Φ is a ground state of H. ■

4.2 Renormalized Nelson Hamiltonian

4.2.1 Definition of Renormalized Nelson Hamiltonian

In this section we consider two-body Nelson Hamiltonian with space dimension $d = 3$, but one of two matters has the infinite mass and a cutoff function of the scalar field operator is removed. The Hamiltonian acts in $L^2(\mathbb{R}^3_x \times \mathbb{R}^3_y) \otimes \mathscr{F}$. It reads

$$\left(-\frac{1}{2m_0}\Delta_y - \frac{1}{2m}\Delta_x\right) + H_\mathrm{f} + g\phi_y + g\phi_x. \tag{4.14}$$

Here we introduce a coupling constant $g \in \mathbb{R}$. The Hamiltonian (4.14) is translation invariant since no external potentials are introduced. Here

$$\phi_x = \int_{\mathbb{R}^3_x\times\mathbb{R}^3_y}^{\oplus} \phi(\rho(x))dxdy, \quad \phi_y = \int_{\mathbb{R}^3_x\times\mathbb{R}^3_y}^{\oplus} \phi(\rho(y))dxdy.$$

Then the ground state of (4.14) does not exists for arbitrary g. So let us suppose $m_0 = \infty$ and take the new coordinate with $y = 0$. Then the Hamiltonian is

$$H_\mathrm{N} = -\frac{1}{2m}\Delta_x + H_\mathrm{f} + g\phi_0 + g\phi_x,$$

acting on $\mathscr{H} = L^2(\mathbb{R}^3) \otimes \mathscr{F}$.

Assumption 4.17 (*Cutoff functions*) We set

$$\hat{\varphi}(k) = \begin{cases} 0, & |k| < \kappa, \\ (2\pi)^{-3/2}, & \kappa \le |k| \le \lambda, \\ 0, & |k| > \lambda. \end{cases}$$

λ is an ultraviolet cutoff parameter and κ an infrared cutoff parameter.

In spirit one would like to have $\varphi(x) = \delta(x)$. I.e., $\kappa = 0$ and $\lambda = \infty$. In this case unfortunately H_N is both infrared and ultraviolet divergent: If $\varphi(x) \to \delta(x)$ then the ground state energy tends to $-\infty$. These divergences are however of a mild nature and can be unraveled through the Gross transformation.

Definition 4.18 (*Gross transformation*) Let $\kappa > 0$. Unitary operator e^T is called Gross transformation, where $T = \int_{\mathbb{R}^3}^{\oplus} T(x)dx$ with

$$T(x) = -g \int \frac{\hat{\varphi}(k)}{\sqrt{2\omega(k)}} \left\{ \left(\beta(k) e^{ikx} + \frac{1}{\omega(k)} \right) a(k) - \left(\beta(k) e^{-ikx} + \frac{1}{\omega(k)} \right) a^{\dagger}(k) \right\} dk,$$

and $\beta(x)$ is a propagator given by

$$\beta(k) = \frac{1}{\omega(k) + \frac{|k|^2}{2m}}.$$

In the case of $\kappa = 0$ we can not define Gross transformation since $\hat{\varphi}/\omega^{3/2} \notin L^2(\mathbb{R}^3)$. Note that $\beta \in L^2(\mathbb{R}^3)$. Define

$$A = A_{\kappa\lambda} = g \int k \frac{\hat{\varphi}(k)}{\sqrt{2\omega(k)}} \beta(k) a(k) e^{ikx} dk.$$

Hence it is obtained that

$$e^T H_{\mathrm{N}} e^{-T} = H_{\mathrm{el}} + H_{\mathrm{f}} + \frac{1}{m}(pA + A^* p) + \frac{1}{2m}\left(A^2 + A^{*2} + 2A^* A\right) + \Delta V + R.$$

Here $p = -i\nabla$ and

$$H_{\mathrm{el}} = -\frac{1}{2m}\Delta - \frac{g^2}{4\pi}\frac{1}{|x|},$$

$$\Delta V = g^2 \left(\frac{1}{4\pi |x|} - \int \frac{|\hat{\varphi}(k)|^2}{\omega(k)^2} e^{ikx} dk \right),$$

$$R = -g^2 \int_{\mathbb{R}^3} \frac{|\hat{\varphi}(k)|^2}{2\omega(k)} \left(\frac{1}{\omega(k)} + \beta(k) \right) dk.$$

If the limit $\kappa \to 0$ the transformation e^T is no longer unitarily implemented and one changes to a distinct Fock representation. It is seen that there exists $\alpha(\kappa, \lambda) > 0$ such that

$$\left| \frac{1}{4\pi |x|} - \int_{\mathbb{R}^3} \frac{|\hat{\varphi}(k)|^2}{\omega(k)^2} e^{ikx} dk \right| \le \frac{\alpha(\kappa, \lambda)}{4\pi |x|}$$

for $x \neq 0$ with $\lim_{(\kappa,\lambda)\to(0,\infty)} \alpha(\kappa, \lambda) = 0$. Furthermore R can be treated as renormalization terms which diverges to $-\infty$ as $\lambda \to \infty$. Then we remover ΔV and R from $e^T H_{\mathrm{N}} e^{-T}$ and we obtain the definition of the Hamiltonian considered in this section.

Definition 4.19 (*Gross transformed Nelson Hamiltonian*) Let $0 \le \kappa < \lambda < \infty$. Gross transformed Nelson Hamiltonian $H_{\kappa\lambda}$ is defined by

$$H_{\kappa\lambda} = H_{\mathrm{el}} + H_{\mathrm{f}} + \frac{1}{m}(pA + A^* p) + \frac{1}{2m}\left(A^2 + A^{*2} + 2A^* A\right).$$

Remark 4.20 $H_{\kappa\lambda}$ is defined for $0 \le \kappa < \lambda < \infty$. If $0 < \kappa < \lambda < \infty$, $H_{\kappa\lambda}$ is unitary equivalent to H_{N}, but if $\kappa = 0$, $H_{\kappa\lambda}$ is not unitary equivalent to H_{N}. In the case of $\kappa = 0$, H_{N} has no ground state but as is seen below $H_{0\lambda}$ has the ground state.

We set

$$V(x) = -\frac{g^2}{4\pi}\frac{1}{|x|}.$$

Thus the spectrum of the hydrogen like atom H_{el} is established as

$$\mathrm{sp}(H_{\mathrm{el}}) = \left\{-\frac{1}{2}\left(\frac{g^2}{4\pi}\right)^2\frac{1}{n^2}\right\}_{n=1}^{\infty} \cup [0,\infty)$$

and $0 \notin \mathrm{sp_p}(H_{\mathrm{el}})$. In particular

$$E_{\mathrm{el}} = \inf \mathrm{sp}(H_{\mathrm{el}}) = -\frac{g^4}{32\pi^2}$$

and E_{el} is simple. Let

$$H_0 = (H_{\mathrm{el}} - E_{\mathrm{el}}) + H_{\mathrm{f}}.$$

4.2.2 Removal of Both Infrared and Ultraviolet Cutoffs

We can remove both cutoffs κ and λ imposed on $H_{\kappa\lambda}$. Nelson [49] proved the following proposition.

Proposition 4.21 (Nelson [49]) *Let $t > 0$. Then in the operator norm it follows that*

$$\lim_{\lambda\to\infty} e^{-tH_{\kappa\lambda}} = e^{-tH_{\kappa\infty}},$$
$$\lim_{\kappa\to 0} e^{-tH_{\kappa\infty}} = e^{-tH_{0\infty}}.$$

$H_{0\infty}$ is called the renormalized Nelson Hamiltonian.

Remark 4.22 Nelson proved Proposition 4.21 in the strong topology, but it can be shown in the uniform operator topology. See [46].

The cutoff function imposed on A is $k\hat{\varphi}e^{-ikx}\beta/\sqrt{\omega}$. Set $\hat{\varphi} = 1$. We have $k_\mu e^{-ikx}\beta/\sqrt{\omega} \notin L^2(\mathbb{R}^3)$. Precisely

$$\mathbb{1}_{|k|<1}\frac{k_\mu e^{-ikx}\beta}{\sqrt{\omega}} \in L^2(\mathbb{R}^3) \quad \text{but} \quad \mathbb{1}_{|k|\ge 1}\frac{k_\mu e^{-ikx}\beta}{\sqrt{\omega}} \notin L^2(\mathbb{R}^3)$$

implies that $A_{0\infty}$ can not be defined, but

$$\frac{k_\mu e^{-ikx}\beta}{\omega^{\frac{1}{2}+\varepsilon}} \in L^2(\mathbb{R}^3), \quad \varepsilon > 0,$$

suggests that $A_{0\infty}$ can be defined on $D(H_{\mathrm{f}}^{1/2})$ by the inequality:

$$\|a(f)\Psi\| \le \|f/\sqrt{\omega}\|\|H_{\mathrm{f}}^{1/2}\Psi\|.$$

On the other hand

$$\|a^\dagger(f)\Psi\| \le \|f/\sqrt{\omega}\|\|H_{\mathrm{f}}^{1/2}\Psi\| + \|f\|\|\Psi\|$$

suggests however that $A_{0\infty}^*$ can not be defined on $D(H_{\mathrm{f}}^{1/2})$. Let us prepare some useful inequalities to define $H_{0\infty}$.

Lemma 4.23 *Let $f, g \in L^2(\mathbb{R}^3)$. Then*

$$\|(H_{\mathrm{f}}+\mathbb{1})^{-1/2}a(f)a(g)(H_{\mathrm{f}}+\mathbb{1})^{-1/2}\| \le A + B,$$

where

$$A = (\|g_1/\sqrt{\omega}\| + \|g_2/\sqrt{\omega}\|)(\|f_1/\sqrt{\omega}\| + \|f_1\|) + (\|g_1/\sqrt{\omega}\| + \|g_1\|)\|f_2/\sqrt{\omega}\|,$$
$$B = \sqrt{3}\|g_2/\omega^{1/4}\|\|f_2/\omega^{1/4}\|$$

with $f_1 = f\mathbb{1}_{|k|<1}$, $f_2 = f\mathbb{1}_{|k|\ge 1}$, $g_1 = g\mathbb{1}_{|k|<1}$ and $g_2 = g\mathbb{1}_{|k|\ge 1}$.

Proof Let $S = \{k \in \mathbb{R}^3 | |k| < 1\}$. Define $\mathcal{F}_1 = \mathscr{F}(L^2(S))$ and $\mathcal{F}_2 = \mathscr{F}(L^2(S^c))$. By the unitary operator $\mathscr{U}_{\mathrm{c}} : \mathscr{F} \longrightarrow \mathcal{F}_1 \otimes \mathcal{F}_2$,

$$\mathscr{U}_{\mathrm{c}} a^\sharp(f)\mathscr{U}_{\mathrm{c}}^{-1} = a^\sharp(f_1)\otimes \mathbb{1} + \mathbb{1}\otimes a^\sharp(f_2), \quad \mathscr{U}_{\mathrm{c}} H_{\mathrm{f}} \mathscr{U}_{\mathrm{c}}^{-1} = H_{\mathrm{f}1}\otimes \mathbb{1} + \mathbb{1}\otimes H_{\mathrm{f}2}.$$

Here $H_{\mathrm{f}1} = \mathrm{d}\Gamma(\omega\lceil_{|k|<1})$ and $H_{\mathrm{f}2} = \mathrm{d}\Gamma(\omega\lceil_{|k|\ge 1})$. Set $a^\sharp(f_1)\otimes \mathbb{1} = a_1^\sharp(f)$ and $\mathbb{1}\otimes a^\sharp(f_2) = a_2^\sharp(f)$. We have

$$a(f)a(g) = a_1(f)a_1(g) + a_1(f)a_2(g) + a_2(f)a_1(g) + a_2(f)a_2(g).$$

We estimate above four terms separately. We have

$$|(\phi, a_1(f)a_1(g)\psi)| \le (\|f_1/\sqrt{\omega}\| + \|f_1\|)\|g_1/\sqrt{\omega}\|\|(H_{\mathrm{f}1}+\mathbb{1})^{1/2}\phi\|\|(H_{\mathrm{f}1}+\mathbb{1})^{1/2}\psi\|.$$

It follows that

$$\|(H_{\mathrm{f}}+\mathbb{1})^{-1/2}a_1(f)a_1(g)(H_{\mathrm{f}1}+\mathbb{1})^{-1/2}\| \le (\|f_1/\sqrt{\omega}\| + \|f_1\|)\|g_1/\sqrt{\omega}\|.$$

For the second term we have

$$|(\phi, a_1(f)a_2(g)\psi)| \le (\|f_1/\sqrt{\omega}\| + \|f_1\|)\|g_2/\sqrt{\omega}\|\|(H_{f1} + \mathbb{1})^{1/2}\phi\|\|(H_{f2} + \mathbb{1})^{1/2}\psi\|.$$

Then we have

$$\|(H_f + \mathbb{1})^{-1/2}a_1(f)a_2(g)(H_f + \mathbb{1})^{-1/2}\| \le (\|f_1/\sqrt{\omega}\| + \|f_1\|)\|g_2/\sqrt{\omega}\|.$$

The third term is estimated in the similar way to the second and we have

$$\|(H_f + \mathbb{1})^{-1/2}a_2(f)a_1(g)(H_f + \mathbb{1})^{-1/2}\| \le (\|g_1/\sqrt{\omega}\| + \|g_1\|)\|f_2/\sqrt{\omega}\|.$$

Finally we have

$$\|(H_f + \mathbb{1})^{-1/2}a_2(f)a_2(g)(H_f + \mathbb{1})^{-1/2}\| \le \sqrt{3}\|f_2/\sqrt{\omega}\|\|g_2/\sqrt{\omega}\|.$$

This is the hard part and we refer to see [29, Lemma A.2(ii)]. ■

We emphasize that the right-hand side of the inequality in Lemma 4.23 does *not* contain $\|f_2\|$ and $\|g_2\|$.

By Lemma 4.23 $(\phi, A^2\psi)$, $(A^2\phi, \psi)$ and $(A\phi, A\psi)$ are well defined for $\phi, \psi \in D(H_f^{1/2})$. The symmetric forms B_0 and $B_{A\kappa\lambda}$ for $\kappa \ge 0$ and $\lambda \le \infty$

$$\begin{aligned}
B_0(\phi, \psi) &= (H_0^{1/2}\phi, H_0^{1/2}\psi),\\
B_{A\kappa\lambda}(\phi, \psi) &= \frac{1}{m}(p\phi, A\psi) + \frac{1}{m}(A\phi, p\psi) + (\phi, V\psi) + E_{el}(\phi, \psi)\\
&\quad + \frac{1}{2m}(\phi, A^2\psi) + \frac{1}{2m}(A^2\phi, \psi) + \frac{1}{m}(A\phi, A\psi)
\end{aligned}$$

are well defined on $D(H_0^{1/2}) \times D(H_0^{1/2})$.

We introduce an abstract theorem.

Theorem 4.24 (KLMN theorem) *Let A be a positive self-adjoint operator and q_A the quadratic form associated with A. Suppose that β is a symmetric quadratic form on $Q(q_A)$ such that $|\beta(f, f)| \le aq_A(f, f) + b(f, f)$ with some $a < 1$ and $b \ge 0$. Then there exists a unique self-adjoint operator C such that $Q(q_C) = Q(q_A)$ and $q_C(f, g) = q_A(f, g) + \beta(f, g)$ for $f, g \in Q(q_C)$. Furthermore C is bounded below by $-b$ and any domain of essential self-adjointness for A is a form core for C.*

Proof See [51]. ■

By Lemma 4.23 we can see that there exist positive constants a_1 and a_2 independent of κ, λ, and g such that with $C(g) = g^2a_1 + |g|a_2$

$$|B_{A\kappa\lambda}(\phi, \phi)| \le C(g)\|(H_0 + \mathbb{1})^{1/2}\phi\|^2 + |E_{el}|\|\phi\|^2, \tag{4.15}$$

Define the quadratic form

$$B_{\kappa\lambda} = B_0 + B_{A\kappa\lambda}.$$

Suppose that $C(g) < 1$. KLMN theorem implies that there exists a self-adjoint operator $K_{\kappa\lambda}$ associated with $B_{\kappa\lambda}$.

Proposition 4.25 (Nelson [49]) *Let* $0 \le \kappa$ *and* $\lambda \le \infty$. *Suppose that* $C(g) < 1$. *Then* $K_{\kappa\lambda} = H_{\kappa\lambda}$.

Proof See [49] and [29, Proposition 3.3]. ∎

Remark 4.26 The limit Hamiltonian $H_{0\infty}$ is formally denoted by

$$H_{0\infty} = H_{\mathrm{el}} + H_{\mathrm{f}} + \frac{1}{m}(pA_{0\infty} + A^*_{0\infty}p) + \frac{1}{2m}(A^2_{0\infty} + A^{*2}_{0\infty} + 2A^*_{0\infty}A_{0\infty}),$$

where

$$A_{0\infty} = g\int k\frac{1}{\sqrt{2\omega(k)}}\beta(k)a(k)e^{ikx}dk.$$

Since $\||k|\beta/\sqrt{\omega}\| = \infty$, $A^*_{0\infty}$ is not defined on $D(H_0^{1/2})$ as an operator. $H_{0\infty}$ is however well defined as the form on the domain $D(H_0^{1/2}) \times D(H_0^{1/2})$.

4.2.3 Spatial Localization and Pull-Through Formula

Let $E_{\kappa\lambda} = \inf \mathrm{sp}(H_{\kappa\lambda})$ and $E_{\kappa\lambda,0} = \inf \mathrm{sp}(H_{\kappa\lambda,V=0})$ be the ground state energy of $H_{\kappa\lambda}$ with $V = 0$. Since $V \le 0$, it follows that

$$E_{\kappa\lambda,0} - E_{\kappa\lambda} \ge 0.$$

For $(\chi, \tilde{\chi}) \in l_{\mathrm{loc}}(\mathbb{R}^3)$ we define $\chi_R(x) = \chi(x/R)$ and $\tilde{\chi}_R(x) = \tilde{\chi}(x/R)$ with parameter $R > 0$. Let $0 \le \kappa < \lambda < \infty$. We define $E_{\kappa\lambda,R}$ by

$$E_{\kappa\lambda,R} = \inf_{F\in H_0^{1/2}, \tilde{\chi}_R F\neq 0} \frac{B_{\kappa\lambda}(\tilde{\chi}_R F, \tilde{\chi}_R F)}{(\tilde{\chi}_R F, \tilde{\chi}_R F)}.$$

The ionization energy is given by $\Sigma_{\kappa\lambda} = \lim_{R\to\infty} E_{\kappa\lambda,R}$.

Lemma 4.27 (Griesmer, Lieb and Loss [21]) *Suppose* $C(g) < 1$. *Then*

$$\inf_{0<\kappa<\lambda<\infty}(E_{\kappa\lambda,0} - E_{\kappa\lambda}) > -E_{\mathrm{el}} > 0.$$

Proof Let f be the ground state of H_{el}, i.e., $H_{\mathrm{el}}f = E_{\mathrm{el}}f$. Set K for $H_{\kappa\lambda}$ with $V = 0$. Take arbitrary $\varepsilon > 0$. Let $F \in D(K)$ be such that $(F, KF) < E_{\kappa\lambda,0} + \varepsilon$. We have

$$(fF, (H_{\kappa\lambda} - (E_{\kappa\lambda,0} + \varepsilon + E_{\rm el})fF)$$
$$= (fF, f(K - (E_{\kappa\lambda,0} + \varepsilon))F) + (fF, (-i\nabla f)(-i\nabla + A + A^*)F).$$

Since $(fF, (-i\nabla f)(A + A^*)F) = 0$, we have

$$\begin{aligned}
&(fF, (H_{\kappa\lambda} - (E_{\kappa\lambda,0} + \varepsilon + E_{\rm el}))fF)\\
&= \int_{\mathbb{R}^3} \left((F, KF)_{\mathscr{F}}(x) - (E_{\kappa\lambda,0} + \varepsilon)\|F\|^2_{\mathscr{F}}(x)\right) f(x)^2 dx\\
&+ \int_{\mathbb{R}^3} (F, -i\nabla F)_{\mathscr{F}}(x) f(x)(-i\nabla f)(x)dx.
\end{aligned}$$

Since K is translation invariant there exists $F_y \in D(K)$ such that

$$(F_y, KF_y)_{\mathscr{F}}(x) = (F, KF)_{\mathscr{F}}(x + y), \quad \|F_y\|^2_{\mathscr{F}}(x) = \|F\|^2_{\mathscr{F}}(x + y).$$

Thus inserting F_y to F we have

$$\begin{aligned}
\Omega_y =& (fF_y, (H_{\kappa\lambda} - (E_{\kappa\lambda,0} + \varepsilon + E_{\rm el}))fF_y)\\
=& \int_{\mathbb{R}^3} (F, (K - (E_{\kappa\lambda,0} + \varepsilon)F)_{\mathscr{F}}(x) f(x + y)^2 dx\\
&+ \int_{\mathbb{R}^3} (F, -i\nabla F)_{\mathscr{F}}(x) f(x - y)(-i\nabla f)(x - y)dx
\end{aligned}$$

and since

$$\int_{\mathbb{R}^3} dy \int_{\mathbb{R}^3} (F, -i\nabla F)_{\mathscr{F}}(x) f(x - y)(-i\nabla f)(x - y)dx = 0,$$

it follows that

$$\int_{\mathbb{R}^3} \Omega_y dy = \int_{\mathbb{R}^3} dy \int_{\mathbb{R}^3} (F, K - (E_{\kappa\lambda,0} + \varepsilon)F)_{\mathscr{F}}(x) f(x + y)^2 dx < 0.$$

This implies that there exists $y_0 \in \mathbb{R}^3$ such that $\Omega_{y_0} < 0$. Thus we conclude that $E_{\kappa\lambda,0} - E_{\kappa\lambda} \geq -E_{\rm el} - \varepsilon$ and

$$E_{\kappa\lambda,0} - E_{\kappa\lambda} \geq -E_{\rm el} > 0$$

follows. ■

Let E_O be the spectral projection of $H_{\kappa\lambda}$ for measurable set $O \subset \mathbb{R}$.

Lemma 4.28 *Suppose $C(g) < 1$. Let $\beta = g^2/8\pi$. Then there exists $\Lambda > 0$ such that*

$$\sup_{0<\kappa<\lambda<\infty} \|e^{\beta|x|} E_{(-\infty,\Lambda]}\| < \infty.$$

Proof If $\Lambda \geq E_{\kappa\lambda}$ and $\beta > 0$ satisfies $\Lambda + \beta^2 < \Sigma_{\kappa\lambda}$, $\|e^{\beta|x|}E_{(-\infty,\Lambda]}\| < \infty$ by Proposition 3.17. Note that $E_{\kappa\lambda,0} \leq \Sigma_{\kappa\lambda}$ by Corollary 3.12, and $-E_{\mathrm{el}} \leq E_{\kappa\lambda,0} - E_{\kappa\lambda}$ by Lemma 4.27. Take $\beta^2 = -E_{\mathrm{el}}/2 = g^4/64\pi^2$ and $\Lambda = E_{\kappa\lambda}$. We have

$$\Lambda + \beta^2 = E_{\kappa\lambda} + \frac{g^4}{64\pi^2} < E_{\kappa\lambda,0} \leq \Sigma_{\kappa\lambda}.$$

The lemma is proven. ■

Since $H_{\kappa\lambda}$ uniformly converges to $H_{0\infty}$ in the sense of semigroup, $E_{\kappa\lambda}$ converges to $\inf \mathrm{sp}(H_{0\infty})$ as $\lambda \to \infty$ and then $\kappa \to 0$. There exist κ_*, λ_* and g_* such that for all $\kappa < \kappa_*$ and $\lambda > \lambda_*$ and $|g| < g_*$

$$-E_{\min} \leq E_{\kappa\lambda} \leq -E_{\max}$$

with some positive constants $E_{\min}$ and $E_{\max}$ independent of κ, λ and g.

Assumption 4.29 Suppose $C(g) < 1$ and the triplet (κ, λ, g) is such that $\kappa < \kappa_*$ and $\lambda > \lambda_*$ and $|g| < g_*$ but $g \neq 0$.

Suppose Assumption 4.29. From (4.15) it follows that

$$\bar{a}(H_{\kappa\lambda} - E_{\kappa\lambda}) - \bar{b} \leq H_0 \leq a(H_{\kappa\lambda} - E_{\kappa\lambda}) + b, \tag{4.16}$$

where

$$\bar{a} = \frac{1}{(1 + C(g))}, \quad \bar{b} = \frac{C(g) - E_{\mathrm{el}} + E_{\min}}{1 + C(g)},$$

$$a = \frac{1}{(1 - C(g))}, \quad b = \frac{C(g) - E_{\mathrm{el}} - E_{\max}}{1 - C(g)}.$$

Corollary 4.30 *Suppose Assumption 4.29. Then there exist constants C_1 and $C_2(\epsilon)$ independent of κ, λ and g such that*

$$\|(H_0 + \mathbb{1})^{1/2}(H_{\kappa\lambda} - E_{\kappa\lambda} + \epsilon)^{-1/2}\| \leq \frac{C_1}{\sqrt{\epsilon}}, \quad \epsilon > 0,$$

$$\|(H_{\kappa\lambda} - E_{\kappa\lambda} + \epsilon)^{1/2}(H_0 + \mathbb{1})^{-1/2}\| \leq C_2(\epsilon), \quad \epsilon \geq 0.$$

Proof By (4.16) we have

$$\|(H_0 + \mathbb{1})^{1/2}(H_{\kappa\lambda} - E_{\kappa\lambda} + \epsilon)^{-1/2}\| \leq \left(\frac{1 + b}{\epsilon}\right)^{1/2},$$

$$\|(H_{\kappa\lambda} - E_{\kappa\lambda} + \epsilon)^{1/2}(H_0 + \mathbb{1})^{-1/2}\| \leq \left(\frac{\bar{a}\epsilon + \bar{b}}{\bar{a}}\right)^{1/2}.$$

The corollary follows. ■

Suppose $\lambda < \infty$ and $\kappa > 0$. We denote by $\Psi_{\kappa\lambda}$ a ground state of $H_{\kappa\lambda}$. Let

$$h(k) = \frac{k\beta(k)e^{-ikx}\hat{\varphi}(k)}{\sqrt{\omega(k)}}.$$

We have bounds

$$|h(k)| \le \begin{cases} \frac{|\hat{\varphi}(k)|}{\sqrt{\omega(k)}} & |k| < 1, \\ \frac{|\hat{\varphi}(k)|}{|k|\sqrt{\omega(k)}} & |k| \ge 1. \end{cases}$$

These bounds will be used to estimate $\|N^{1/2}\Psi_{\kappa\lambda}\|$. The pull-through formula we have

$$\|N^{1/2}\Psi_{\kappa\lambda}\|^2 = g^2\int_{\mathbb{R}^3} \|(H_{\kappa\lambda} - E_{\kappa\lambda} + \omega(k))^{-1}(p + A + A^*)h(k)\Psi_{\kappa\lambda}\|^2 dk.$$

We separate $\|N^{1/2}\Psi_{\kappa\lambda}\|^2$ into the infrared region $\int_{|k|<1} \|a(\Psi_{\kappa\lambda})(k)\|^2 dk$ and the ultraviolet region $\int_{|k|\ge 1} \|a(\Psi_{\kappa\lambda})(k)\|^2 dk$:

$$\|N^{1/2}\Psi_{\kappa\lambda}\|^2 = \int_{|k|<1} \|a(\Psi_{\kappa\lambda})(k)\|^2 dk + \int_{|k|\ge 1} \|a(\Psi_{\kappa\lambda})(k)\|^2 dk.$$

Lemma 4.31 (Ultraviolet region) *Suppose Assumption 4.29. Then there exists a constant $c_1 > 0$ independent of κ, λ and g such that*

$$\int_{|k|\ge 1} \|a(\Psi_{\kappa\lambda})(k)\|^2 dk \le g^2 c_1^2 \int_{|k|\ge 1} \frac{|h(k)|^2}{\omega(k)} dk \|\Psi_{\kappa\lambda}\|^2$$

Proof Set $\bar{H}_{\kappa\lambda} = H_{\kappa\lambda} - E_{\kappa\lambda}$. Since

$$\|(\bar{H}_{\kappa\lambda} + \omega(k))^{-1}\| \le \frac{1}{\sqrt{\omega(k)}}\|(\bar{H}_{\kappa\lambda} + \omega(k))^{-1/2}\|,$$

it is sufficient to prove that

$$\sup_{|k|\ge 1} \|(\bar{H}_{\kappa\lambda} + \omega(k))^{-1/2}(p_\mu + A_\mu + A_\mu^*)\Psi_{\kappa\lambda}\| \le c_1\|\Psi_{\kappa\lambda}\|.$$

We have by Corollary 4.30

$$\begin{aligned} &\|(\bar{H}_{\kappa\lambda} + \omega(k))^{-1/2} p_\mu \Psi_{\kappa\lambda}\| \\ &\le \|(\bar{H}_{\kappa\lambda} + \omega(k))^{-1/2}(H_0 + \mathbb{1})^{1/2}\| \|(H_0 + \mathbb{1})^{-1/2} p_\mu\| \|\Psi_{\kappa\lambda}\| \le \frac{C_1}{\sqrt{\omega(k)}}\|\Psi_{\kappa\lambda}\|. \end{aligned}$$

Since

$$\begin{aligned}
\|(H_0+\mathbb{1})^{-1/2}A_\mu\Psi_{\kappa\lambda}\|^2 &\le \|A_\mu\Psi_{\kappa\lambda}\|^2 \le \|\frac{\beta}{\sqrt{2}}\|^2\|H_{\mathrm{f}}^{1/2}\Psi_{\kappa\lambda}\|^2\\
&\le \|\frac{\beta}{\sqrt{2}}\|^2(\Psi_{\kappa\lambda}, H_0\Psi_{\kappa\lambda}) \le \|\frac{\beta}{\sqrt{2}}\|^2 b\|\Psi_{\kappa\lambda}\|,
\end{aligned}$$

where b is in (4.16), we have

$$\begin{aligned}
&\|(\bar{H}_{\kappa\lambda}+\omega(k))^{-1/2}A_\mu\Psi_{\kappa\lambda}\|\\
&\le \|(\bar{H}_{\kappa\lambda}+\omega(k))^{-1/2}(H_0+\mathbb{1})^{1/2}\|\|(H_0+\mathbb{1})^{-1/2}A_\mu\Psi_{\kappa\lambda}\| \le \frac{C_1\|\frac{\beta}{\sqrt{2}}\|}{\sqrt{\omega(k)}}\sqrt{b}\|\Psi_{\kappa\lambda}\|
\end{aligned}$$

and

$$\begin{aligned}
&\|(\bar{H}_{\kappa\lambda}+\omega(k))^{-1/2}A_\mu^*\Psi_{\kappa\lambda}\|\\
&\le \|(\bar{H}_{\kappa\lambda}+\omega(k))^{-1/2}(H_0+\mathbb{1})^{1/2}\|\|(H_0+\mathbb{1})^{-1/2}A_\mu^*\Psi_{\kappa\lambda}\| \le \frac{C_1\|\frac{\beta}{\sqrt{2}}\|}{\sqrt{\omega(k)}}\|\Psi_{\kappa\lambda}\|.
\end{aligned}$$

It follows that

$$\|(\bar{H}_{\kappa\lambda}+\omega(k))^{1/2}(p_\mu+A_\mu+A_\mu^*)\Psi_{\kappa\lambda}\| \le \frac{C_1}{\sqrt{\omega(k)}}\left(1+(\sqrt{b}+1)\|\frac{\beta}{\sqrt{2}}\|\right)\|\Psi_{\kappa\lambda}\|.$$

The lemma follows. ■

Lemma 4.32 (Infrared region) *Suppose Assumption 4.29. Then there exists a constant c_2 independent of κ, λ and g such that*

$$\int_{|k|<1}\|a(\Psi_{\kappa\lambda})(k)\|^2 dk \le g^2 c_2\||x|\Psi_{\kappa\lambda}\|^2.$$

Proof We have

$$[H_{\kappa\lambda}, e^{-ikx}x] = [H_{\kappa\lambda}, e^{-ikx}]x + e^{-ikx}[H_{\kappa\lambda}, x].$$

A direct calculation shows that

$$\begin{aligned}
&[H_{\kappa\lambda}, x] = -i(p+A+A^*),\\
&[H_{\kappa\lambda}, e^{-ikx}] = -(p+A+A^*)ke^{-ikx} + \frac{1}{2}|k|^2 e^{-ikx}.
\end{aligned}$$

We obtain that

$$h(k)(p+A+A^*)$$
$$= -i(H_{\kappa\lambda}-E_{\kappa\lambda})h(k)xe^{-ikx} - i(p+A+A^*)kh(k)xe^{-ikx} + \frac{i}{2}|k|^2h(k)xe^{-ikx}.$$

The pull-through formula yields that

$$\int_{|k|<1} \|a(\Psi_{\kappa\lambda})(k)\|^2 dk \le a+b+c,$$

where

$$a = g^2 \int_{|k|<1} \|(H_{\kappa\lambda}-E_{\kappa\lambda}+\omega(k))^{-1}(H_{\kappa\lambda}-E_{\kappa\lambda})h(k)xe^{-ikx}\Psi_{\kappa\lambda}\|^2 dk,$$
$$b = \frac{g^2}{4}\int_{|k|<1} \|(H_{\kappa\lambda}-E_{\kappa\lambda}+\omega(k))^{-1}|k|^2h(k)xe^{-ikx}\Psi_{\kappa\lambda}\|^2 dk,$$
$$c = g^2 \int_{|k|<1} \|(H_{\kappa\lambda}-E_{\kappa\lambda}+\omega(k))^{-1}(p+A+A^*)kh(k)xe^{-ikx}\Psi_{\kappa\lambda}\|^2 dk.$$

We have

$$a \le g^2 \int_{|k|<1} |h(k)|^2 dk \||x|\Psi_{\kappa\lambda}\|^2, \quad b \le \frac{g^2}{4}\int_{|k|<1}\left(\frac{|k|^2|h(k)|}{\omega(k)}\right)^2 dk\||x|\Psi_{\kappa\lambda}\|^2.$$

From Lemma 4.31 it follows that

$$c \le g^2 C_1^2 \left(1+(\sqrt{b}+1)\|\frac{\beta}{\sqrt{2}}\|\right)^2 \int_{|k|<1}\left|\frac{kh(k)}{\omega(k)}\right|^2 dk\||x|\Psi_{\kappa\lambda}\|^2.$$

Then we get the desired results. ■

Theorem 4.33 *Suppose Assumption 4.29. Then there exists a constant C independent of κ, λ and g such that*

$$\frac{\|N^{1/2}\Psi_{\kappa\lambda}\|}{\|\langle x\rangle\Psi_{\kappa\lambda}\|^2} \le g^2 C\left\{\int_{|k|\ge 1}\frac{|\hat{\varphi}(k)|^2}{\omega(k)^4}dk + \int_{|k|<1}\left(\omega(k)+\frac{1}{\omega(k)}\right)|\hat{\varphi}(k)|^2 dk\right\}. \quad (4.17)$$

In particular there exists $C^* > 0$ independent of $0 < \kappa$ and $\lambda < \infty$ such that

$$\sup_{0<\kappa<\lambda<\infty} \frac{\|N^{1/2}\Psi_{\kappa\lambda}\|}{\|\Psi_{\kappa\lambda}\|} \le g^2 C^*. \quad (4.18)$$

Proof Equation (4.17) follows from Lemmas 4.32 and 4.31. The integral on the right-hand side of (4.17) is nondecreasing as $\kappa \downarrow 0$ and $\lambda \uparrow \infty$. We have

$$\lim_{\kappa\downarrow 0,\lambda\uparrow\infty}\int_{|k|\geq 1}\frac{|\hat{\varphi}(k)|^2}{\omega(k)^4}dk+\int_{|k|<1}\left(\omega(k)+\frac{1}{\omega(k)}\right)|\hat{\varphi}(k)|^2dk$$
$$=\frac{1}{(2\pi)^3}\int_{|k|\geq 1}\frac{1}{\omega(k)^4}dk+\frac{1}{(2\pi)^3}\int_{|k|<1}\left(\omega(k)+\frac{1}{\omega(k)}\right)dk<\infty.$$

By Lemma 4.28 we have $\sup_{0<\kappa<\lambda<\infty}\|e^{\beta|x|}\Psi_{\kappa\lambda}\|/\|\Psi_{\kappa\lambda}\|<\infty$. Then we proved (4.18). ∎

4.2.4 Existence of the Ground State Without Cutoffs

In this section we show the existence of a ground state of $H_{\kappa\lambda}$ with $\lambda=\infty$ and $\kappa=0$. I.e., $H_{0\infty}$ has a ground state.

Let $P_{\rm el}$ be the projection onto the interval $[E_{\rm el},-\epsilon]$ of the spectrum of $H_{\rm el}$, where we suppose $\epsilon>0$ and $E_{\max}-\epsilon>0$. Let $P=P_{\rm el}\otimes P_\Omega$ and $Q=P_{\rm el}^\perp\otimes P_\Omega$.

Lemma 4.34 *Suppose Assumption 4.29. Then there exist constants D_1 and D_2 independent of κ, λ and g such that*

$$(\Psi_{\kappa\lambda},Q\Psi_{\kappa\lambda})\leq D(g)\|\Psi_{\kappa\lambda}\|^2 \tag{4.19}$$

with $D(g)=|g|D_1+g^2D_2$.

Proof Since $(Q\Psi_{\kappa\lambda},(H_{\kappa\lambda}-E_{\kappa\lambda})\Psi_{\kappa\lambda})=0$, it follows that

$$(Q\Psi_{\kappa\lambda},(H_{\rm el}+H_{\rm f}-E_{\kappa\lambda})\Psi_{\kappa\lambda})=-(Q\Psi_{\kappa\lambda},H_A\Psi_{\kappa\lambda}), \tag{4.20}$$

where $H_A=pA+A^*p+\frac{1}{2}(A^2+2A^*A+A^{*2})$. The left-hand side of (4.20) can be estimated as

$$(-\epsilon+E_{\max})\|Q\Psi_{\kappa\lambda}\|^2\leq(-\epsilon-E_{\kappa\lambda})(Q\Psi_{\kappa\lambda},\Psi_{\kappa\lambda})\leq(Q\Psi_{\kappa\lambda},(H_{\rm el}+H_{\rm f}-E_{\kappa\lambda})\Psi_{\kappa\lambda}).$$

We shall estimate the right-hand side. Let $-(Q\Psi_{\kappa\lambda},(H_A+V)\Psi_{\kappa\lambda})=a+b$, where

$$a=-(Q\Psi_{\kappa\lambda},(pA+A^*p)\Psi_{\kappa\lambda}),\quad b=-\frac{1}{2}(Q\Psi_{\kappa\lambda},A^2+2A^*A+A^{*2}\Psi_{\kappa\lambda}).$$

We note that $[(H_0+\mathbb{1})^{1/2},Q_{\kappa\lambda}]=0$. We have

$$\begin{aligned}&|(Q\Psi_{\kappa\lambda},pA\Psi_{\kappa\lambda})|\\&\leq|g|\|(H_0+\mathbb{1})^{1/2}Q\Psi_{\kappa\lambda}\|\|(H_0+\mathbb{1})^{-1/2}p\|\|A(H_0+\mathbb{1})^{-1/2}\|\|(H_0+\mathbb{1})^{1/2}\Psi_{\kappa\lambda}\|\\&\leq\|\frac{\beta}{\sqrt{2}}\|\|(H_0+\mathbb{1})^{1/2}p\|\|(H_0+\mathbb{1})^{1/2}\Psi_{\kappa\lambda}\|^2.\end{aligned}$$

Similarly we have

$$|(Q\Psi_{\kappa\lambda}, A^* p\Psi_{\kappa\lambda})| \le |g| \| \frac{\beta}{\sqrt{2}} \| \| (H_0 + \mathbb{1})^{-1/2} p \| \| (H_0 + \mathbb{1})^{1/2} \Psi_{\kappa\lambda} \|^2.$$

From (4.16) it follows that $\|(H_0 + \mathbb{1})^{1/2}\Psi_{\kappa\lambda}\|^2 \le |g| C \|\Psi_{\kappa\lambda}\|^2$ with some constant C independent of κ, λ and g. Then $|a| \le |g| D_1' \|\Psi_{\kappa\lambda}\|^2$ follows. It is obtained by Lemma 4.23 that

$$(Q\Psi_{\kappa\lambda}, A^2\Psi_{\kappa\lambda})| \le g^2 D \|(H_0 + \mathbb{1})^{1/2}\Psi_{\kappa\lambda}\|^2,$$
$$(Q\Psi_{\kappa\lambda}, A^{*2}\Psi_{\kappa\lambda})| \le g^2 D \|(H_0 + \mathbb{1})^{1/2}\Psi_{\kappa\lambda}\|^2$$

with some constant D independent of κ, λ and g. Moreover

$$(Q\Psi_{\kappa\lambda}, A^* A\Psi_{\kappa\lambda})| \le g^2 \| \frac{\beta}{\sqrt{2}} \|^2 \|\Psi_{\kappa\lambda}\|^2.$$

Hence we have $|b| \le g^2 D_2' \|\Psi_{\kappa\lambda}\|^2$. Then the lemma follows. ■

We define

$$g_c = \sup\left\{ g \in \mathbb{R} \,|\, |g| < g_*, C(g) < 1 \text{ and } g^2 C^* + D(g) < 1 \right\}.$$

Here $C(g)$ is given by (4.15), $D(g)$ by (4.19) and C^* by (4.18).

Theorem 4.35 (Hirokawa, Hiroshima and Spohn [29, Theorem 1.2]) *Suppose* $|g| < g_c$ *but* $g \ne 0$ *with some* g_c. *Then* $H_{0\infty}$ *has a ground state* $\psi_{0\infty}$ *such that*

$$\psi_{0\infty} \in D(e^{c|x|} \otimes N^{1/2}) \quad 0 \le \forall c \le \frac{g^2}{8\pi}.$$

Proof Since $P \ge \mathbb{1} - Q - N$, by Theorem 4.33 and Lemma 4.34 we have

$$(\Psi_{\kappa\lambda}, P\Psi_{\kappa\lambda}) \ge \left(1 - g^2 C^* - D(g)\right) \|\Psi_{\kappa\lambda}\|^2 \tag{4.21}$$

The right-hand side of (4.21) is strictly positive. We normalize $\Psi_{\kappa\lambda}$ and take a subsequence λ such that the weak limit $\psi_{\kappa,\infty}$ of $\psi_{\kappa,\lambda'}$ exists as $\lambda' \to \infty$. Since P is a finite rank operator, by (4.21), $(\psi_{\kappa,\infty}, P\psi_{\kappa,\infty})$ does not vanish, which implies also $\psi_{\kappa,\infty} \ne 0$. It is a ground state of $H_{\kappa,\infty}$ and

$$(\psi_{\kappa,\infty}, P\psi_{\kappa,\infty}) \ge 1 - g^2 C^* - D(g) > 0 \tag{4.22}$$

also holds. Once again we take a subsequence κ' such that $\psi_{\kappa',\infty}$ has a weak limit $\psi_{0\infty}$ as $\kappa' \to 0$. Take $k' \to 0$ on the both sides of (4.22). We have

$$(\psi_{0,\infty}, P\psi_{0,\infty}) \ge 1 - g^2 C^* - D(g) > 0.$$

By this we conclude that $\psi_{0\infty}$ does not vanish. Then $\psi_{0\infty}$ is a ground state of $H_{0\infty}$. ■

The uniqueness of the ground state of $H_{0\infty}$ is proven in Matte and Møller [46] by showing that the semigroup generated by the renormalized Nelson Hamiltonian is positivity improving. Recently the existence of the ground state is also proven under weaker conditions.

Theorem 4.36 (Matte and Møller [46]) *The ground state of $H_{0\infty}$ is unique.*

By using a functional integration Hiroshima and Matte [36] prove the existence of the ground state of both (1) renormalized Nelson Hamiltonian *with* infrared cutoff and (2) renormalized Gross transformed Nelson Hamiltonian *without* infrared cutoff, for arbitrary values of coupling constants. Furthermore the absence of the ground state of renormalized Nelson Hamiltonian *without* infrared cutoff is also shown. In [36] a hypercontractivity and Kolmogorov–Riesz–Fréchet theorem (Theorem 2.13) are used for the proof. Furthermore various localization properties of the ground state are shown.

Chapter 5
Spin-Boson Model

Abstract In this chapter a spin-boson model is studied by path measures. This model describes a two-level atom interacting with a scalar quantum field. The existence of the ground state of the spin-boson model is proven.

5.1 Spin-Boson Hamiltonian

We consider a two-state system linearly coupled to a scaler quantum field. Suppose that the space dimension is d. Consider the Hilbert space

$$\mathscr{H} = \mathbb{C}^2 \otimes \mathscr{F}.$$

Here $\mathscr{F} = \mathscr{F}_{\rm b}(L^2(\mathbb{R}^d))$ is the boson Fock space over $L^2(\mathbb{R}^d)$. We begin by defining the spin-boson Hamiltonian as a selfadjoint operator acting in $\mathscr{H}$. The free field Hamiltonian for spin-boson model is given by

$$H_{\rm f} = {\rm d}\Gamma(\omega),$$

where $\omega = \omega(k) = |k|$. Let

$$\phi_\varphi = \frac{1}{\sqrt{2}}\left(a^\dagger(\hat{\varphi}/\sqrt{\omega}) + a(\tilde{\hat{\varphi}}/\sqrt{\omega})\right).$$

We introduce an assumption on $\hat{\varphi}$ which is supposed in this chapter unless otherwise stated.

Assumption 5.1 (*Cutoff functions*) $\varphi \in \mathscr{S}'(\mathbb{R}^d)$ satisfies that (1) $\hat{\varphi} \in L^1_{\rm loc}(\mathbb{R}^d)$, (2) $\overline{\hat{\varphi}(k)} = \hat{\varphi}(-k)$, (3) $\hat{\varphi}/\omega, \hat{\varphi}/\sqrt{\omega} \in L^2(\mathbb{R}^d)$.

Let σ_x, σ_y and σ_z be the 2×2 Pauli matrices given by

$$\sigma_x = \begin{pmatrix} 0 & 1 \\ 1 & 0 \end{pmatrix}, \quad \sigma_y = \begin{pmatrix} 0 & -i \\ i & 0 \end{pmatrix}, \quad \sigma_z = \begin{pmatrix} 1 & 0 \\ 0 & -1 \end{pmatrix}.$$

F. Hiroshima, *Ground States of Quantum Field Models*,
SpringerBriefs in Mathematical Physics,
https://doi.org/10.1007/978-981-32-9305-2_5

Definition 5.2 (*Spin-boson Hamiltonian*) A spin-boson Hamiltonian is defined by

$$H_{SB} = \varepsilon\sigma_z \otimes \mathbb{1} + \mathbb{1} \otimes H_f + \sigma_x \otimes \phi_\varphi$$

acting on $\mathscr{H}$, where $\varepsilon \geq 0$ is a parameter.

Here 2ε describes the energy gap between the ground state energy and the first excited state energy of the two-level atom. In the same way as the Nelson Hamiltonian we can show the selfadjointness of H_{SB}.

Proposition 5.3 (Selfadjointness) *Spin-boson Hamiltonian H_{SB} is selfadjoint and bounded from below on $D(H_f)$ and essentially selfadjoint on any core of H_f.*

The existence of the ground state can be proven in a similar way (rather easier) to that of the Nelson Hamiltonian by the method of [17], but we demonstrate here to prove it by a stochastic method due to Hirokawa, Hiroshima and Lörinczi [28]. We derive Feynman–Kac type formulas for semigroups generated by a spin-boson Hamiltonian, but we do not intend a comprehensive review of Feynman–Kac type formula, and give the only outline of it.

5.2 Existence of the Ground State

In this section we show the existence of the ground state of a spin-boson Hamiltonian by an application of a Feynman–Kac type formula. To construct the Feynman–Kac type formula of $e^{-tH_{SB}}$ we need to transform H_{SB} to some selfadjoint operator $\tilde{H}_{SB}$.

The rotation group in $\mathbb{R}^3$ has an adjoint representation on $SU(2)$. Let $n \in \mathbb{R}^3$ be a unit vector and $\theta \in [0, 2\pi)$. Thus $e^{(i/2)\theta n\cdot\sigma}$ satisfies that

$$e^{(i/2)\theta n\cdot\sigma}\sigma_\mu e^{-(i/2)\theta n\cdot\sigma} = (R\sigma)_\mu,$$

where R denotes the 3×3 matrix representing the rotation around n with angle θ, and $\sigma = (\sigma_x, \sigma_y, \sigma_z)$. In particular, for $n = (0, 1, 0)$ and $\theta = \pi/2$, we have

$$e^{(i/2)\theta n\cdot\sigma}\sigma_x e^{-(i/2)\theta n\cdot\sigma} = \sigma_z, \quad e^{(i/2)\theta n\cdot\sigma}\sigma_z e^{-(i/2)\theta n\cdot\sigma} = -\sigma_x.$$

Let $U = \exp\left(i\frac{\pi}{4}\sigma_y\right) \otimes \mathbb{1}$. H_{SB} is then transformed as

$$\tilde{H}_{SB} = UH_{SB}U^* = -\varepsilon\sigma_x \otimes \mathbb{1} + \mathbb{1} \otimes H_f + \sigma_z \otimes \phi_\varphi.$$

$\tilde{H}_{SB}$ is realized as

$$\tilde{H}_{SB} = \begin{pmatrix} H_f + \phi_\varphi & -\varepsilon \\ -\varepsilon & H_f - \phi_\varphi \end{pmatrix}.$$

Instead of H_{SB} we consider $\tilde{H}_{SB}$ in what follows. We reset $\tilde{H}_{SB}$ as H_{SB}. Put

$$I = \int_{\mathbb{R}^d} \frac{|\hat{\varphi}(k)|^2}{\omega(k)^3} dk.$$

Whenever $\varepsilon = 0$ the Hamiltonian $H_{\rm SB}$ is diagonal, i.e., we have

$$H_{\rm SB} = \begin{pmatrix} H_{\rm f} + \phi_\varphi & 0 \\ 0 & H_{\rm f} - \phi_\varphi \end{pmatrix}.$$

It is known that $H_{\rm f} + \phi_\varphi$ is called van Hove–Miyatake Hamiltonian and has the unique ground state if and only if $I < \infty$ ([4, Chap. 13]), which implies that $H_{\rm SB}$ with $\varepsilon = 0$ has a two-fold degenerate ground state if and only if $I < \infty$.

Next we consider the case of $\varepsilon \neq 0$ and $\Omega_{\rm SB} = \begin{pmatrix} 1 \\ 1 \end{pmatrix} \otimes \Omega$ and $E = \inf \sigma(H_{\rm SB})$. Write

$$\Psi_{\rm g}^T = e^{-T(H_{\rm SB}-E)}\Omega_{\rm SB}, \quad T \geq 0,$$

and

$$\gamma(T) = (\Omega_{\rm SB}, \frac{\Psi_{\rm g}^T}{\|\Psi_{\rm g}^T\|})^2 = \frac{(\Omega_{\rm SB}, e^{-TH_{\rm SB}}\Omega_{\rm SB})^2}{(\Omega_{\rm SB}, e^{-2TH_{\rm SB}}\Omega_{\rm SB})}.$$

The next criterion is useful for showing the existence of a ground state of $H_{\rm SB}$.

Proposition 5.4 *Suppose that* $\lim_{T\to\infty} \gamma(T) = a$. *If* $a > 0$, *then* $H_{\rm SB}$ *has the ground state.*

Proof Since $a > 0$, for sufficiently large T it follows that $b < \sqrt{\gamma(T)}$ with some $b > 0$. Let $E(\cdot)$ be the spectral projection of $H_{\rm SB}$ associated with vector $\Omega_{\rm SB}$. Let $\delta = \inf \mathrm{supp} E(\cdot)$. Then $\delta = E$ and we put $f(c) = E([\delta, \delta + c))$. It is enough to show that $f(0) \neq 0$. We have

$$\sqrt{\gamma(T)} = \frac{\int_{[\delta,\delta+c)} e^{-T(\lambda-\delta)} dE + \int_{[\delta+c,\infty)} e^{-T(\lambda-\delta)} dE}{\left(\int_{[\delta,\infty)} e^{-2T(\lambda-\delta)} dE\right)^{1/2}}.$$

By Schwarz inequality on the numerator we have

$$\frac{\int_{[\delta,\delta+c)} e^{-T(\lambda-\delta)} dE}{\left(\int_{[\delta,\infty)} e^{-2T(\lambda-\delta)} dE\right)^{1/2}} \leq \sqrt{f(c)}.$$

Since we have bounds:

$$(1) \int_{[\delta+c,\infty)} e^{-T(\lambda-\delta)}dE \leq e^{-Tc},$$

$$(2) \int_{[\delta,\infty)} e^{-2T(\lambda-\delta)}dE \geq \int_{[\delta,\delta+c/2)} e^{-2T(\lambda-\delta)}dE \geq e^{-cT} f(c/2),$$

we can see that

$$\frac{\int_{[\delta+c,\infty)} e^{-T(\lambda-\delta)}dE}{\left(\int_{[\delta,\infty)} e^{-2T(\lambda-\delta)}dE\right)^{1/2}} \leq e^{-Tc/2}/\sqrt{f(c/2)}.$$

Then it follows that

$$\sqrt{\gamma(T)} \leq \sqrt{f(c)} + \frac{e^{-Tc/2}}{\sqrt{f(c/2)}}.$$

Take $T \to \infty$ on both sides above. Then we obtain $b \leq f(c)$ for any $c > 0$, and $b \leq f(0)$ can be derived by taking the limit of $c \downarrow 0$. Hence $0 \neq f(0)$ and we can conclude that a ground state of H_{SB} exists. ■

A ground state of H_{SB} exists if $\lim_{T\to\infty} \gamma(T) > 0$ by Propositions 5.4. To see this we can represent $\gamma(T)$ by path measures.

Let $\mathbb{Z}_2 = \{-1, 1\}$. Let $(N_t)_{t\in\mathbb{R}}$ be the two-sided Poisson process[1] with unit intensity on a probability space (S, B_S, P). Define a $\mathbb{Z}_2$-valued stochastic process $(\sigma_t)_{t\in\mathbb{R}}$ by $\sigma_t = \sigma(-1)^{N_t}$ for $t \in \mathbb{R}$, where $\sigma \in \mathbb{Z}_2$. I.e., $\sigma_0 = \sigma$. We define

$$W(X, t) = \frac{1}{2}X \int_{\mathbb{R}^d} e^{-|t|\omega(k)} \frac{|\hat{\varphi}(k)|^2}{\omega(k)^2}dk, \quad (X, t) \in \mathbb{Z}_2 \times \mathbb{R}$$

We can prove the crucial lemma.

Lemma 5.5 *It follows that*

$$\|\Psi_g^T\|^2 = e^{2TE}e^{2\varepsilon T}\sum_{\sigma\in\mathbb{Z}_2} \mathbb{E}\left[e^{\frac{1}{2}\int_{-T}^T ds \int_{-T}^T dt W(\sigma_{\varepsilon s}\sigma_{\varepsilon t}, s-t)}\right],$$

$$(\Omega_{\mathrm{SB}}, \Psi_g^T) = e^{TE}e^{\varepsilon T}\sum_{\sigma\in\mathbb{Z}_2} \mathbb{E}\left[e^{\frac{1}{2}\int_0^T ds \int_0^T dt W(\sigma_{\varepsilon s}\sigma_{\varepsilon t}, s-t)}\right].$$

Here $\mathbb{E}[\ldots] = \int_S \ldots dP$.

Proof See [28, Corollary 1.4]. ■

---

[1]For each $t \geq 0$, N_t is a non-negative integer-valued random variable such that $\mathbb{E}[N_t = n] = e^{-t}t^n/n!$

We can see that

$$\left| \int_{-T}^{0} ds \int_{0}^{T} dt W(\sigma_{\varepsilon s}\sigma_{\varepsilon t}, s-t) \right| \le \frac{1}{2} I \tag{5.1}$$

uniformly in T and in paths.

Theorem 5.6 (Hirokawa, Hiroshima and Lőrinczi [28, Theorem 2.4])
Suppose $I < \infty$. Then H_{SB} has the ground state.

Proof Simply set $W = W(\sigma_{\varepsilon s}\sigma_{\varepsilon t}, s-t)$. We decompose the integral as

$$\int_{-T}^{T} ds \int_{-T}^{T} dt W = \int_{-T}^{0} ds \int_{-T}^{0} dt W + \int_{0}^{T} ds \int_{0}^{T} dt W + 2 \int_{-T}^{0} ds \int_{0}^{T} dt W.$$

By (5.1) we obtain

$$\|\Psi_{\mathrm{g}}^{T}\|^2 \le e^{2TE} e^{2\varepsilon T} \sum_{\sigma \in \mathbb{Z}_2} \mathbb{E}\left[e^{\frac{1}{2}(\int_{-T}^{0} ds \int_{-T}^{0} dt W + \int_{0}^{T} ds \int_{0}^{T} dt W + I)} \right].$$

By the independence of N_t and N_{-s} for $-s < 0 < t$, and N_t and N_{-t} have the same law we furthermore obtain that

$$\begin{aligned}
\|\Psi_{\mathrm{g}}^{T}\|^2 &\le e^{2TE} e^{2\varepsilon T} \sum_{\sigma \in \mathbb{Z}_2} \left(\mathbb{E}\left[e^{\frac{1}{2}\int_{0}^{T} ds \int_{0}^{T} dt W(\sigma_{\varepsilon s}\sigma_{\varepsilon t}, s-t)} \right] \right)^2 e^{\frac{1}{2}I} \\
&\le \left(e^{TE} e^{\varepsilon T} \sum_{\sigma \in \mathbb{Z}_2} \mathbb{E}\left[e^{\frac{1}{2}\int_{0}^{T} ds \int_{0}^{T} dt W(\sigma_{\varepsilon s}\sigma_{\varepsilon t}, s-t)} \right] \right)^2 e^{\frac{1}{2}I} = (\Omega_{\mathrm{SB}}, \Psi_{\mathrm{g}}^{T})^2 e^{\frac{1}{2}I}.
\end{aligned}$$

Hence $\gamma(T) \ge e^{-\frac{1}{2}I}$ and $\lim_{T\to\infty} \gamma(T) \ge e^{-\frac{1}{2}I} > 0$. Then the ground state of H_{SB} exists. ■

Chapter 6
Enhanced Bindings

Abstract In this chapter we discuss enhanced bindings for the Pauli–Fierz Hamiltonian with dipole approximation and for the N-body Nelson Hamiltonian. It is shown that an interaction with a quantum field enhances binding and sufficiently strong couplings consequently produce the ground state. It is also shown that the Pauli–Fierz Hamiltonian with dipole approximation with sufficiently weak coupling constants has no ground state if it has no ground state whenever a coupling constant is zero.

6.1 Enhanced Binding for the Pauli–Fierz Model

6.1.1 The Pauli–Fierz Hamiltonian with Dipole Approximation

The existence of the ground state for the Pauli–Fierz Hamiltonian $H_{\rm PF}$ can be regarded as a stability property of the spectrum. In Chap. 3 we assume that $H_{\rm PF}$ has a ground state for $\alpha = 0$, which amounts to the existence of a ground state for particle Hamiltonian $H_{\rm p}$. Then we can prove that $H_{\rm PF}$ admits a ground state for any $\alpha \neq 0$. In contrast, in this section we assume that $H_{\rm PF}$ has no ground state for $\alpha = 0$. In fact, this will be the case for a sufficiently shallow negative external potential V because of the d-dimensional Lieb-Thirring inequality [42, 44, 47]:

$$N \leq a_d \int_{\mathbb{R}^d} |V(x)|^{d/2} dx, \quad d \geq 3,$$

where N denotes the number of non-positive eigenvalues of d-dimensional Schrödinger operator $-\frac{1}{2}\Delta + V$ and a_d is a constant depending on only space dimension d. We expect however the interaction between the quantized radiation field and $H_{\rm p}$ enhances binding, and hence a non-binding potential should become a binding potential at sufficiently strong coupling strength. This is precisely our main result in this section, and this phenomenon is called the enhanced binding.

F. Hiroshima, *Ground States of Quantum Field Models*,
SpringerBriefs in Mathematical Physics,
https://doi.org/10.1007/978-981-32-9305-2_6

We take the so-called dipole approximation for the Pauli–Fierz Hamiltonian. In this approximation the quantized radiation field $A(x)$ is replaced by $A(0)$.

Definition 6.1 (*The Pauli–Fierz Hamiltonian with dipole approximation*) The Pauli–Fierz Hamiltonian with dipole approximation $H_{\rm PF}^{\rm dip}$ is defined by

$$H_{\rm PF}^{\rm dip} = \frac{1}{2m}(-i\nabla - \alpha A(0))^2 + H_{\rm rad} + V,$$

where we revive the mass of the electron m and a coupling constant $\alpha \in \mathbb{R}$.

We introduce some technical assumptions on $\hat{\varphi}$ to diagonalize $H_{\rm PF}^{\rm dip}$.

Assumption 6.2 (Cutoff functions) Let $\varphi \in \mathscr{S}'(\mathbb{R}^3)$ and $\hat{\varphi} \in L^1_{\rm loc}(\mathbb{R}^3)$. We suppose (1), (2), (4) or (1), (3), (4):

(1) $\sqrt{\omega}\hat{\varphi}, \hat{\varphi}/\omega \in L^2(\mathbb{R}^3)$, $\overline{\hat{\varphi}(k)} = \hat{\varphi}(-k)$ and $\hat{\varphi}$ is rotation invariant.
(2) $\hat{\varphi}(k) \neq 0$ for $k \neq 0$, and $\rho(s) = |\hat{\varphi}(\sqrt{s})|^2 s^{1/2} \in L^\varepsilon([0,\infty), ds)$ for some $1 < \varepsilon$, and there exists $0 < C < 1$ such that $|\rho(s+h) - \rho(s)| \leq K|h|^C$ for all s and $0 \leq h \leq 1$,
(3) $\hat{\varphi}(k) = 0$ on $\{k||k| < \lambda\} \cup \{k||k| > \Lambda\}$ and $\hat{\varphi}(k) \neq 0$ on $\{k|\lambda \leq |k| \leq \Lambda\}$ with some $\Lambda > \lambda > 0$,
(4) $\|\hat{\varphi}\|_\infty < \infty$ and $\|\hat{\varphi}\omega\|_\infty < \infty$.

We assume Assumption 6.2 throughout this section.

Proposition 6.3 *Let $V \in V_{\rm Kato}$ and we suppose that $\int_{\mathbb{R}^3} |\hat{\varphi}(k)|^2/\omega(k)^3 dk < \infty$. Then there exists a unitary operator $\mathscr{U}$ such that $\mathscr{U}$ maps $D(-\Delta) \cap D(H_{\rm rad})$ onto itself and*

$$\mathscr{U}^{-1} H_{\rm PF}^{\rm dip} \mathscr{U} = h_{\rm eff} + H_{\rm rad} + \delta V + g,$$

where $h_{\rm eff}$ denotes the effective Hamiltonian given by

$$h_{\rm eff} = -\frac{1}{2m_{\rm eff}}\Delta + V,$$

and error term δV is of the form: $\delta V = T^{-1}VT - V$ with

$$T = \exp(-K \cdot \nabla),$$

$$K_\mu = \frac{\alpha}{\sqrt{2}} \sum_{j=1,2} \int \frac{e_\mu(k,j)}{\sqrt{\omega(k)}} \left(\frac{\hat{\varphi}(k)}{m_{\rm eff}(k)\omega(k)} a^\dagger(k,j) + \frac{\hat{\varphi}(-k)}{m_{\rm eff}(k)\omega(k)} a(k,j) \right) dk.$$

Here the function $m_{\rm eff}(k)$ is given by $m_{\rm eff}(k) = D_+(\omega(k)^2)$ with

$$D_\pm(s) = m - \frac{4\pi}{3}\alpha^2 \left(\lim_{\epsilon \downarrow 0} \int_{|s-x|>\epsilon} \frac{\rho(|x|)}{s-x} dx \mp \pi i \rho(s) \right).$$

Furthermore the effective mass $m_{\rm eff}$ is given by

$$m_{\rm eff} = m + \frac{2}{3}\alpha^2 \|\hat{\varphi}/\omega\|^2,$$

and the additional constant g by

$$g = \frac{3}{2\pi}\int_{-\infty}^{\infty} \frac{\alpha^2 \frac{2}{3}\left\| \frac{t\hat{\varphi}}{t^2+\omega^2}\right\|^2}{m + \alpha^2\frac{2}{3}\left\|\frac{\hat{\varphi}}{\sqrt{t^2+\omega^2}}\right\|^2} dt.$$

Proof See [2] and [38, Theorem 3.5]. ∎

The physical reasoning behind the enhanced binding for the Pauli–Fierz Hamiltonian is simple. As the particle binds photons it acquires an effective mass $m_{\rm eff} = m_{\rm eff}(\alpha^2)$ which is increasing in $|\alpha|$. Let us start with

$$H_{\rm p} = -\frac{1}{2m}\Delta + V,$$

where $V(x) \leq 0$, $\lim_{|x|\to\infty} V(x) = 0$, such that $H_{\rm p}$ has only the absolutely continuous spectrum $[0, \infty)$. Proposition 6.3 claims that $H_{\rm PF}^{\rm dip} \cong h_{\rm eff} + H_{\rm rad} + g + \delta V$. Note that $m_{\rm eff}$ rather than m appears in the transformed Hamiltonian. Proposition 6.3 roughly says that $H_{\rm PF}^{\rm dip}$ may be replaced by

$$h_{\rm eff} = -\frac{1}{2m_{\rm eff}}\Delta + V,$$

which binds for sufficiently strong α. $h_{\rm eff}$ has the ground state for large $|\alpha|$ but has no ground state for small $|\alpha|$. We can prove that for $|\alpha| > \alpha_*$ with some α_*, $H_{\rm PF}^{\rm dip}$ has the ground state. See Fig. 1.5. The key ingredient to show the existence of the ground state is to take a scaling. The dilated $H_{\rm p}$,

$$H_{\rm p}(\kappa) = -\frac{\kappa^2}{2m}\Delta + V(\frac{x}{\kappa}),$$

has the same spectrum as that of $H_{\rm p}$. We couple to the boson field as

$$\kappa^2\left\{\frac{1}{2m}(-i\nabla - \alpha A)^2 + \frac{1}{\kappa^2}V(\frac{x}{\kappa}) + H_{\rm rad}\right\} \cong \frac{1}{2m}(-i\nabla - \kappa\alpha A)^2 + V + \kappa^2 H_{\rm rad} = H_\kappa$$

and prove that,

$$s-\lim_{\kappa\to\infty}\left(H_\kappa-\kappa^2\alpha^2 g+z\right)^{-1}=\left(-\frac{1}{2m_{\rm eff}}\Delta+V+z\right)^{-1}\otimes P_{\rm f} \tag{6.1}$$

with $\mathrm{Im} z \neq 0$ and $P_{\rm f}$ the projection onto the ground state of $H_{\rm rad}+\frac{\alpha^2}{2m}A^2(0)$. See [3, 33] for the proof of (6.1). For sufficiently large α the limiting Hamiltonian has a ground state. If we can prove that this ground state persists for large κ, we are done. In other words $H_{\rm PF}^{\rm dip}$ with the external potential $\frac{1}{\kappa^2}V(\frac{x}{\kappa})$, $\kappa \gg 1$, has a ground state for large α.

We are also interested in studying the spectrum of $H_{\rm PF}^{\rm dip}$ for sufficiently small α but $\alpha \neq 0$. We prove that $H_{\rm PF}^{\rm dip}$ has no ground state in sufficiently weak coupling to the quantized radiation field. The Birman-Schwinger principle can be applied to $H_{\rm PF}^{\rm dip}$ and show the absence of the ground state of $H_{\rm PF}^{\rm dip}$ for sufficiently small $|\alpha|$.

6.1.2 *Absence of the Ground State for Small* $|\alpha|$

In this section it is shown that the Pauli–Fierz Hamiltonian with dipole approximation has no ground state for sufficiently small $|\alpha|$.

The unbinding for the Schrödinger operator $H_{\rm p}=-\frac{1}{2m}\Delta+V$ is proven by the Birman-Schwinger principle. Formally one has

$$H_{\rm p}=\frac{1}{2m}(-\Delta)^{1/2}\left(\mathbb{1}+2m(-\Delta)^{-1/2}V(-\Delta)^{-1/2}\right)(-\Delta)^{1/2}.$$

If m is sufficiently small, then $2m(-\Delta)^{-1/2}V(-\Delta)^{-1/2}$ is a strict contraction operator. Hence the operator $\mathbb{1}+2m(-\Delta)^{-1/2}V(-\Delta)^{-1/2}$ has a bounded inverse and $H_{\rm p}$ has no eigenvalue in $(-\infty, 0]$. More precisely the Birman-Schwinger principle states that

$$\mathrm{dimRan}\,\mathbb{1}_{[\frac{1}{2m},\infty)}(V^{1/2}(-\Delta)^{-1}V^{1/2})\geq \mathrm{dimRan}\,\mathbb{1}_{(-\infty,0]}(H_{\rm p}). \tag{6.2}$$

For small m the left-hand side equals 0 and thus $H_{\rm p}$ has no eigenvalues in $(-\infty, 0]$. Our approach will be to generalize (6.2) to the Pauli–Fierz Hamiltonian with dipole approximation. We already see that $H_{\rm PF}^{\rm dip}$ can be transformed by $\mathscr{U}$ and one arrives at

$$\mathscr{U}^{-1}H_{\rm PF}^{\rm dip}\mathscr{U}=-\frac{1}{2m_{\rm eff}}\Delta+H_{\rm rad}+W+g$$

involving the effective mass $m_{\rm eff}$ of the dressed particle, the transformed interaction $W=T^{-1}VT$, and the global energy shift g.

Let $h_0=-\frac{1}{2}\Delta$. We assume that $V\in L^1_{\rm loc}(\mathbb{R}^3)$ and V is relatively form-bounded with respect to h_0 with relative bound $a<1$, i.e., $D(|V|^{1/2})\supset D(h_0^{1/2})$ and

$$\||V|^{1/2}\varphi\|^2 \le a\|h_0^{1/2}\varphi\|^2 + b\|\varphi\|^2, \quad \varphi \in D(h_0^{1/2}), \tag{6.3}$$

with some $b > 0$. Under (6.3) Birman-Schwinger operator $R_E = (h_0 - E)^{-1/2}\,|V|^{1/2}$ for $E < 0$ are densely defined. From (6.3) it follows that $R_E^* = |V|^{1/2}(h_0 - E)^{-1/2}$ is bounded and thus R_E is closable. We denote its closure by the same symbol. Let

$$K_E = R_E^* R_E.$$

K_E $(E < 0)$ is a bounded, positive selfadjoint operator and it holds

$$K_E f = |V|^{1/2}\,(h_0 - E)^{-1}\,|V|^{1/2} f, \quad f \in C_0^\infty(\mathbb{R}^3).$$

Now let us consider the case $E = 0$. Let $R_0 = h_0^{-1/2}|V|^{1/2}$. The selfadjoint operator $h_0^{-1/2}$ has the integral kernel[1] $h_0^{-1/2}(x, y) = \dfrac{\sqrt{2\pi}}{|x - y|^2}$. It holds that

$$\left|(h_0^{-1/2}g, |V|^{1/2}f)\right| \le \sqrt{2\pi}\|g\|_2\||V|^{1/2}f\|_{6/5}$$

for $f, g \in C_0^\infty(\mathbb{R}^3)$ by the Hardy-Littlewood-Sobolev inequality.[2] Since $f \in C_0^\infty(\mathbb{R}^3)$ and $V \in L^1_{\rm loc}(\mathbb{R}^3)$, one concludes $\||V|^{1/2}f\|_{6/5} < \infty$. Thus $|V|^{1/2}f \in D(h_0^{-1/2})$ and R_0 is densely defined. Since V is relatively form-bounded with respect to h_0, R_0^* is also densely defined, and R_0 is closable. We denote the closure by the same symbol. We define $K_0 = R_0^* R_0$.

Next let us introduce assumptions on the external potential V.

Assumption 6.4 V satisfies that (1) $V \le 0$ and (2) R_0 is compact.

We recall that the Rollnik class $\mathscr{R}$ of potentials is defined by

$$\mathscr{R} = \left\{ V \left| \int_{\mathbb{R}^3} dx \int_{\mathbb{R}^3} \frac{|V(x)V(y)|}{|x - y|^2} dy < \infty \right.\right\}.$$

By the Hardy-Littlewood-Sobolev inequality, $\mathscr{R} \supset L^p(\mathbb{R}^3) \cap L^r(\mathbb{R}^3)$ with $1/p + 1/r = 4/3$. In particular, $L^{3/2}(\mathbb{R}^3) \subset \mathscr{R}$.

---

[1]The integral kernel of operator $(-\Delta)^{-\alpha/2}$ in d-dimensional space is given by

$$(2\pi)^{-d/2}\frac{c_{d-\alpha}}{c_\alpha}\frac{1}{|x - y|^{d-\alpha}}, \quad c_\nu = \Gamma(\nu/2)2^{\nu/2}, \quad d \ge 3.$$

[2]Let $0 < \lambda < d$ and $1 < p, q$. Suppose that $f \in L^p(\mathbb{R}^d)$ and $g \in L^q(\mathbb{R}^d)$ with $1/p + \lambda/d + 1/q = 2$. Then there exists $C = C(d, \lambda, p)$ such that

$$\left|\int_{\mathbb{R}^d} dx \int_{\mathbb{R}^d} f(x)|x - y|^{-\lambda} g(y) dy\right| \le C\|f\|_p\|g\|_q.$$

Example 6.5 Suppose that V is negative and $V \in \mathscr{R}$. Then $K_0 \in L^2(\mathbb{R}^3 \times \mathbb{R}^3)$. Hence K_0 is Hilbert-Schmidt and Assumption 6.4 is satisfied.

Lemma 6.6 *Suppose Assumption 6.4. Then*

(1) R_E, R_E^* *and* K_E $(E \le 0)$ *are compact.*
(2) Map $E \mapsto \|K_E\|$ *is continuous and monotonously increasing in* $E \le 0$ *and it holds that* $\lim_{E\to-\infty} \|K_E\| = 0$ *and* $\lim_{E\uparrow 0} \|K_E\| = \|K_0\|$.

Proof Under (2) of Assumption 6.4, R_0^* and K_0 are compact. Since $(f, K_E f) \le (f, K_0 f)$ for $f \in C_0^\infty(\mathbb{R}^3)$ can be extended to $f \in L^2(\mathbb{R}^3)$, K_E, R_E and R_E^* are also compact. Thus (1) is proven. It is clear that K_E is monotonously increasing in E. Since R_0 is bounded, $(f, K_E f) \le (f, K_0 f)$ also holds on $L^2(\mathbb{R}^3)$ and

$$K_E = R_0^*(h_0 - E)^{-1} h_0 R_0 \tag{6.4}$$

for $E \le 0$. From this one concludes that $\|K_E - K_{E'}\| \le \|K_0\| \frac{|E-E'|}{|E'|}$ for $E, E' < 0$. Hence $\|K_E\|$ is continuous in $E < 0$. We have to prove the left continuity at $E = 0$. Since $\|K_E\| \le \|K_0\|$ $(E < 0)$, one has $\limsup_{E\uparrow 0} \|K_E\| \le \|K_0\|$. By (6.4) we see that $K_0 = \text{s-}\lim_{E\uparrow 0} K_E$ and

$$\|K_0 f\| = \lim_{E\uparrow 0} \|K_E f\| \le \left(\liminf_{E\uparrow 0} \|K_E\| \right) \|f\|, \quad f \in L^2(\mathbb{R}^3).$$

Hence we have $\|K_0\| \le \liminf_{E\uparrow 0} \|K_E\|$ and $\lim_{E\uparrow 0} \|K_E\| = \|K_0\|$. It remains to prove that $\lim_{E\to-\infty} \|K_E\| = 0$. Since R_0^* is compact, for any $\varepsilon > 0$, there exists a finite rank operator $T_\varepsilon = \sum_{k=1}^n (\varphi_k, \cdot)\psi_k$ such that $n = n(\varepsilon) < \infty$, $\varphi_k, \psi_k \in L^2(\mathbb{R}^3)$ and $\|R_0^* - T_\varepsilon\| < \varepsilon$. It holds that $\|K_E\| \le \left(\varepsilon + \|T_\varepsilon h_0 (h_0 - E)^{-1}\|\right) \|R_0\|$. For any $f \in L^2(\mathbb{R}^3)$, we have $\|T_\varepsilon h_0(h_0 - E)^{-1} f\| \le \left(\sum_{k=1}^n \|h_0(h_0 - E)^{-1}\varphi_k\| \|\psi_k\|\right) \|f\|$ and $\lim_{E\to-\infty} \|T_\varepsilon h_0 (h_0 - E)^{-1}\| = 0$, which completes the proof of (2). ■

By (2) of Lemma 6.6, we have $\lim_{E\to-\infty} \||V|^{1/2}(h_0 - E)^{-1/2}\| = 0$. Therefore V is infinitesimally form bounded with respect to h_0 and H_p is the selfadjoint operator associated with the quadratic form

$$f, g \mapsto \frac{1}{m}(h_0^{1/2} f, h_0^{1/2} g) + (|V|^{1/2} f, |V|^{1/2} g)$$

for $f, g \in D(h_0^{1/2})$. Under (2) of Assumption 6.4, the essential spectrum of H_p coincides with that of $-\frac{1}{2m}\Delta$, hence $\mathrm{sp}_\mathrm{ess}(H_\mathrm{p}) = [0, \infty)$. Next we will estimate the spectrum of H_p contained in $(-\infty, 0]$. Let E_O^T, $O \subset \mathbb{R}$, be the spectral projection of selfadjoint operator T for O, and set $N_O(T) = \dim \operatorname{Ran} E_O^T$. The Birman-Schwinger principle states that

$$(E < 0)\ N_{(-\infty, \frac{E}{m}]}\left(H_{\rm p}\right) = N_{[\frac{1}{m}, \infty)}(K_E),$$

$$(E = 0)\ N_{(-\infty, 0]}\left(H_{\rm p}\right) \leq N_{[\frac{1}{m}, \infty)}(K_0).$$

Definition 6.7 (*Critical mass m_c*) We define $m_{\rm c}$ by $m_{\rm c} = \|K_0\|^{-1}$ and m_ε by $m_\varepsilon = \|K_{-\varepsilon}\|^{-1}$ for $\varepsilon > 0$.

Formally m_c is written as $m_{\rm c} = \left\| |V|^{1/2}(-\Delta)^{-1}|V|^{1/2} \right\|^{-1}$.

Lemma 6.8 *Suppose Assumption 6.4.*

(1) If $m < m_{\rm c}$, then $N_{(-\infty,0]}(H_{\rm p}) = 0$.
(2) If $m > m_{\rm c}$, then $N_{(-\infty,0]}(H_{\rm p}) \geq 1$.

Proof It is immediate to see (1) by the Birman-Schwinger principle. Suppose that $m > m_{\rm c}$. Using the continuity and monotonicity of $E \to \|K_E\|$, we see that there exists $\varepsilon > 0$ such that $m_{\rm c} < \|K_{-\varepsilon}\|^{-1} \leq m$. Since $K_{-\varepsilon}$ is a positive compact operator, $\|K_{-\varepsilon}\| \in {\rm sp_p}(K_{-\varepsilon})$ follows and hence $N_{[\frac{1}{m},\infty)}(K_{-\varepsilon}) \geq 1$. Therefore (2) follows again from the Birman-Schwinger principle. ■

Suppose Assumption 6.4. Since the Fock vacuum Ω is the ground state of $H_{\rm rad}$, $H_{\rm p} + H_{\rm rad}$ has a ground state if and only if $H_{\rm p}$ has a ground state. If $m < m_{\rm c}$, $H_{\rm p}$ has no ground state by Lemma 6.8. Then the zero coupling Hamiltonian $H_{\rm p} + H_{\rm rad}$ has no ground state. We extend the Birman-Schwinger type estimate to the Pauli–Fierz Hamiltonian.

Theorem 6.9 (Hiroshima, Suzuki and Spohn [39, Theorem 3.7]) *Let us suppose Assumptions 6.4. If $m_{\rm eff} < m_{\rm c}$, then $H_{\rm PF}^{\rm dip}$ has no ground state.*

Proof Since g is a constant, we prove the absence of the ground state of operator $h_W = -\Delta/2m_{\rm eff} + H_{\rm rad} + W$. Since V is negative, so is W. Hence

$$\inf {\rm sp}(h_W) \leq \inf {\rm sp}(-\frac{1}{2m_{\rm eff}}\Delta + H_{\rm rad}) = 0.$$

Then it suffices to show that h_W has no eigenvalues in $(-\infty, 0]$. Let $E \in (-\infty, 0]$ and set

$$S_E = |W|^{1/2}(h_W - E)^{-1}|W|^{1/2}, \tag{6.5}$$

where $|W|^{1/2}$ is defined by the functional calculus. We shall prove now that if h_W has eigenvalue $E \in (-\infty, 0]$, S_E has eigenvalue 1. Suppose that $(h_W - E)\varphi = 0$ and $\varphi \neq 0$. $S_E|W|^{1/2}\varphi = |W|^{1/2}\varphi$ holds. Moreover if $|W|^{1/2}\varphi = 0$, $W\varphi = 0$ and hence

$$(h_W - E)\varphi = 0,$$

but $-\Delta/2m_{\rm eff}+H_{\rm rad}$ has no eigenvalue. We conclude $|W|^{1/2}\varphi\neq 0$ and S_E has eigenvalue 1. Hence it is sufficient to see $\|S_E\|<1$ to show that h_W has no eigenvalues in $(-\infty,0]$. Notice that Δ and T commute, and

$$\left\|(-\Delta)^{1/2}\,(-\frac{1}{2m_{\rm eff}}\Delta+H_{\rm rad}-E)^{-1}\,(-\Delta)^{1/2}\right\|\leq 2m_{\rm eff}.$$

We have

$$\|S_E\|\leq\left\||V|^{1/2}\left(-\frac{1}{2m_{\rm eff}}\Delta\right)^{-1/2}\right\|^2=m_{\rm eff}\|K_0\|=\frac{m_{\rm eff}}{m_c}<1$$

and the proof is complete. ∎

6.1.3 Existence of the Ground State for Large $|\alpha|$

In this section we show the enhanced binding for the Pauli–Fierz Hamiltonian with dipole approximation, i.e., it is shown that the Pauli–Fierz Hamiltonian with dipole approximation has the ground state for sufficiently large $|\alpha|$.

Let $\kappa>0$ be a scaling parameter. Recall $\kappa^{-2}H_\kappa\cong\frac{1}{2m}(-i\nabla-\alpha A)^2+V_\kappa+H_{\rm rad}$. Set

$$Q(k)=\alpha\frac{\hat\varphi(k)}{m_{\rm eff}(k)}.$$

We introduce assumptions on the external potential V and cutoff function $\hat\varphi$.

Assumption 6.10 (Cutoff functions and external potentials) The external potential V and the cutoff function $\hat\varphi$ satisfy:

(1) $V\in C^1(\mathbb{R}^3)$ and $\nabla V\in L^\infty(\mathbb{R}^3)$,
(2) $\hat\varphi/\omega^{5/2}\in L^2(\mathbb{R}^3)$,
(3) $\sup_{\alpha\in\mathbb{R}}\|Q/\omega^{n/2}\|<\infty$, $n=3,4,5$.

In the same way as in Sect. 6.1.1 H_κ can be diagonalized as

$$H_\kappa\cong h_{\rm eff}+\kappa^2H_{\rm rad}+\delta V_\kappa+g.$$

Here $\delta V_\kappa(x)=V(x+K/\kappa)-V(x)$. We drop g for instance and reset as

$$H_\kappa=h_{\rm eff}+\kappa^2H_{\rm rad}+\delta V_\kappa.$$

We show that H_κ converges to $h_{\rm eff}$ as $\kappa\to\infty$ in some sense. It suggests that H_κ with sufficiently large κ has a ground state if $h_{\rm eff}$ does.

Definition 6.11 (*Critical coupling constant* α_c) Let $m<m_c$ and $\varepsilon>0$. We define

$$\alpha_\varepsilon = \left(\frac{2}{3}\|\hat{\varphi}/\omega\|^2\right)^{-1/2} \sqrt{m_\varepsilon - m}, \quad \varepsilon > 0,$$

$$\alpha_c = \left(\frac{2}{3}\|\hat{\varphi}/\omega\|^2\right)^{-1/2} \sqrt{m_c - m}.$$

Note that $\alpha_c < \alpha_\varepsilon$ because of $m_\varepsilon > m_c$. Since $\lim_{\varepsilon\downarrow 0} m_\varepsilon = m_c$, it holds that

$$\lim_{\varepsilon\downarrow 0} \alpha_\varepsilon = \alpha_c.$$

We note that for $|\alpha| > \alpha_\varepsilon$, $h_{\rm eff}$ has the ground state with negative ground state energy.

We may prove the existence of the ground state of $H_{\rm PF}^{\rm dip}$ for sufficiently large α in a similar manner to Chaps. 3 and 4, but we prove this in this section by momentum lattice approximation which was taken from Glimm and Jaffe [19] and applied to the Pauli–Fierz Hamiltonian by Bach, Fröhlich and Sigal [9] and to the generalized spin-boson Hamiltonian by Arai and Hirokawa [5]. We begin with showing the existence of the ground state of H_κ^ν for $\nu > 0$, where H_κ^ν is defined by H_κ with $\omega(k)$ replaced by $\sqrt{|k|^2 + \nu^2}$.

Let $\Gamma(l, a)$, $l = (l_1, l_2, l_3) \in \mathbb{Z}^3$, $a > 0$, be the momentum lattice with spacing $1/a$, i.e.,

$$\Gamma(l, a) = [\frac{l_1}{a}, \frac{(l_1+1)}{a}) \times [\frac{l_2}{a}, \frac{(l_2+1)}{a}) \times [\frac{l_3}{a}, \frac{(l_3+1)}{a})$$

and

$$\chi_{\Gamma(l,a)}(k) = \begin{cases} 0, & k \notin \Gamma(l, a), \\ a^{3/2}, & k \in \Gamma(l, a). \end{cases}$$

For $L > 0$ and $a > 0$ we define the momentum-lattice-approximated Hamiltonian by

$$\hat{H}_\kappa^\nu = H_\kappa^\nu(L, a) = h_{\rm eff} + \kappa^2 \hat{H}_{\rm rad}^\nu + \delta \hat{V}_\kappa,$$

where $\hat{H}_{\rm rad}^\nu$ and $\delta \hat{V}_\kappa$ are momentum-lattice-approximated operators given by

$$\hat{H}_{\rm rad}^\nu = H_{{\rm rad},L,a}^\nu = \sum_{j=1,2} \int \left(\sum_{|l|\leq L} \chi_{\Gamma(l,a)}(k)(\omega(l) + \nu) \right) a^\dagger(k, j) a(k, j) dk,$$

$$\delta \hat{V}_\kappa = \delta V_{\kappa,L,a} = V(\cdot + K_{L,a}/\kappa) - V$$

and $K_{L,a} = (K_{L,a,1}, K_{L,a,2}, K_{L,a,3})$ is the column of the field operator defined by

$$K_{L,a,\mu} = \frac{1}{\sqrt{2}} \sum_{j=1,2} \int \sum_{|l|\leq L} \chi_{\Gamma(l,a)}(k)(\rho_\mu(l, j) a^\dagger(k, j) + \overline{\rho_\mu(l, j)} a(k, j)) dk.$$

Here we set $\rho_\mu(k, j) = e_\mu(k, j) Q(k)/\omega(k)^{3/2}$. We can show that

$$\|K_{L,a,\mu}\Psi\| \leq C \left(\left\| \sum_{|l|\leq L} \frac{\chi_{\Gamma(l,a)} Q(l)}{\sqrt{\omega}\omega(l)^{3/2}} \right\| + \left\| \sum_{|l|\leq L} \frac{\chi_{\Gamma(l,a)} Q(l)}{\omega(l)^{3/2}} \right\| \right) \|(\hat{H}^{\nu}_{\mathrm{rad}} + \mathbb{1})^{1/2}\Psi\|$$

with some constant C. It follows that $\lim_{L\to\infty}\lim_{a\to\infty} \hat{H}^{\nu}_{\kappa} = H^{\nu}_{\kappa}$ in the uniform resolvent sense. We identify $\ell^2 = \ell^2(\mathbb{Z}^3)$ with the subspace of $L^2(\mathbb{R}^3)$ by

$$\ell^2(\mathbb{Z}^3) \ni \{f(l)\}_{l\in\mathbb{Z}^3} \cong a^{3/2} \sum_{l\in\mathbb{Z}^3} f(l)\chi_{\Gamma(l,a)}(\cdot) \in L^2(\mathbb{R}^3).$$

$L^2(\mathbb{R}^3)$ is decomposed as $L^2(\mathbb{R}^3) = \ell^2 \oplus \ell^{2\perp}$ and the identification

$$\mathcal{H} \cong (\mathcal{H}_a \otimes \mathcal{K}) \oplus \mathcal{H}_a$$

follows. Here $\mathcal{H}_a = L^2(\mathbb{R}^3) \otimes \mathcal{F}(\ell^2)$ and $\mathcal{K} = \oplus_{n=1}^{\infty} \mathcal{F}^{(n)}(\ell^{2\perp})$. In particular we can see that $\mathcal{H}_a^{\perp} \cong \mathcal{H}_a \otimes \mathcal{K}$ and that $\hat{H}^{\nu}_{\kappa}$ is reduced by $\mathcal{H}_a$. We set $K = \hat{H}^{\nu}_{\kappa}\lceil_{\mathcal{H}_a}$ and $K^{\perp} = \hat{H}^{\nu}_{\kappa}\lceil_{\mathcal{H}_a^{\perp}}$. We have

$$\hat{H}^{\nu}_{\kappa} = K^{\perp} \oplus K.$$

We can immediately see that

$$K^{\perp} \cong K \otimes \mathbb{1} + \mathbb{1} \otimes \kappa^2 \hat{H}^{\nu}_{\mathrm{rad}}\lceil_{\mathcal{K}}. \tag{6.6}$$

We denote the ground state energy of selfadjoint operator T by $E(T)$. In view of (6.6) we have $E(K^{\perp}) \geq E(K) + \nu$. In what follows we estimate the spectrum of K.

Lemma 6.12 *Let $\Psi \in D(-\Delta) \cap D(H^{1/2}_{\mathrm{rad}})$. Then*

(1) $\Psi \in D(\delta V_{\kappa})$ and $\|\delta V_{\kappa}\Psi\| \leq \theta_{\kappa}\|(H^{\nu}_{\mathrm{rad}} + \mathbb{1})^{1/2}\Psi\|$, where

$$\theta_{\kappa} = \frac{1}{\kappa} C \|\nabla V\|_{\infty} (\|Q/\omega^2\| + \|Q/\omega^{3/2}\|)$$

with some constant C,

(2) $\Psi \in D(\delta \hat{V}_{\kappa})$ and $\|\delta \hat{V}_{\kappa}\Psi\| \leq \hat{\theta}_{\kappa}\|(\hat{H}^{\nu}_{\mathrm{rad}} + \mathbb{1})^{1/2}\Psi\|$, where

$$\hat{\theta}_{\kappa} = \theta_{\kappa,L,a} = \frac{1}{\kappa} C' \|\nabla V\|_{\infty} \left(\left\| \sum_{|l|\leq L} \frac{\chi_{\Gamma(l,a)} Q(l)}{\sqrt{\omega}\omega(l)^{3/2}} \right\| + \left\| \sum_{|l|\leq L} \frac{\chi_{\Gamma(l,a)} Q(l)}{\omega(l)^{3/2}} \right\| \right)$$

with some constant C'.

Proof We have $\|\delta V_\kappa \Psi\| \leq \frac{1}{\kappa}\|\nabla_\mu V\|_\infty \|K_{\mu,L,a}\Psi\|$. Then (1) follows. (2) is similarly proven. ■

Let $\inf \mathrm{sp}(h_{\mathrm{eff}}) = e_{\mathrm{eff}}$. It follows that $E(H_\kappa^\nu) \leq e_{\mathrm{eff}} + \frac{3\theta_\kappa}{2}$ and $E(\hat{H}_\kappa^\nu) \leq e_{\mathrm{eff}} + \frac{3\hat{\theta}_\kappa}{2}$. We set $\overline{h_{\mathrm{eff}}} = h_{\mathrm{eff}} - e_{\mathrm{eff}}$. Suppose $|\alpha| > \alpha_\varepsilon$. Since $m_{\mathrm{eff}} > m_\varepsilon > m_{\varepsilon/2}$, we see that $e_{\mathrm{eff}} \leq E(-\frac{1}{2m_\varepsilon}\Delta + V) \leq -\frac{\varepsilon}{2m_\varepsilon}$. In particular $|e_{\mathrm{eff}}| > 0$.

Lemma 6.13 *Suppose $|\alpha| > \alpha_\varepsilon$. Let a, L and κ be sufficiently large such that $\min\{|e_{\mathrm{eff}}|/3, 2\kappa^2\} > \hat{\theta}_\kappa$. Then for ν such that $|e_{\mathrm{eff}}| > 3\hat{\theta}_\kappa + \nu$, we have*

$$K - E(K) - \nu \geq E^{\overline{h_{\mathrm{eff}}}}_{[0,|e_{\mathrm{eff}}|)} \otimes \left((\kappa^2 - \frac{\hat{\theta}_\kappa}{2})\hat{H}^\nu_{\mathrm{rad}} - 3\hat{\theta}_\kappa - \nu \right).$$

Proof Set $\bar{K} = K - E(K)$, $X = E^{\overline{h_{\mathrm{eff}}}}_{[0,|e_{\mathrm{eff}}|)}$ and $Y = E^{\overline{h_{\mathrm{eff}}}}_{[|e_{\mathrm{eff}}|,\infty)}$. We directly see that on $\mathscr{H}_a$

$$\begin{aligned}
\bar{K} - \nu &\geq \overline{h_{\mathrm{eff}}} + \delta\hat{V}_\kappa + \kappa^2 \hat{H}^\nu_{\mathrm{rad}} - \frac{3}{2}\hat{\theta}_\kappa - \nu \geq \overline{h_{\mathrm{eff}}} + (\kappa^2 - \frac{\hat{\theta}_\kappa}{2})\hat{H}^\nu_{\mathrm{rad}} - 3\hat{\theta}_\kappa - \nu \\
&\geq |e_{\mathrm{eff}}| Y \otimes \mathbb{1} - \hat{\theta}'_\kappa (X+Y) \otimes \mathbb{1} + (\kappa^2 - \frac{\hat{\theta}_\kappa}{2})(X+Y) \otimes \hat{H}^\nu_{\mathrm{rad}},
\end{aligned}$$

where $\hat{\theta}'_\kappa = 3\hat{\theta}_\kappa + \nu$. Then

$$\bar{K} - \nu \geq (|e_{\mathrm{eff}}| - \hat{\theta}'_\kappa) Y \otimes \mathbb{1} + (\kappa^2 - \frac{\hat{\theta}_\kappa}{2}) Y \otimes \hat{H}^\nu_{\mathrm{rad}} + X \otimes \left((\kappa^2 - \frac{\hat{\theta}_\kappa}{2})\hat{H}^\nu_{\mathrm{rad}} - \hat{\theta}'_\kappa \right).$$

Since $|e_{\mathrm{eff}}| - \hat{\theta}'_\kappa > 0$ and $\kappa^2 - \frac{\hat{\theta}_\kappa}{2} > 0$ by the assumption, we have the lemma. ■

Set $T = K - E(K) - \nu$ as an operator in $\mathscr{H}_a$. Define $\mathscr{H}_a(+) = E^T_{[0,\infty)}\mathscr{H}_a$ and $\mathscr{H}_a(-) = E^T_{[-\nu,0)}\mathscr{H}_a$.

Lemma 6.14 *Suppose $|\alpha| > \alpha_\varepsilon$ and that $\min\{|e_{\mathrm{eff}}|/3, 2\kappa^2\} > \hat{\theta}_\kappa$. Then for ν such that $|e_{\mathrm{eff}}| > 3\hat{\theta}_\kappa + \nu$, $T\lceil_{\mathscr{H}_a(-)}$ has a purely discrete spectrum, i.e.,*

$$\mathrm{sp}(K) \cap [E(K), E(K) + \nu) \subset \mathrm{sp}_{\mathrm{disc}}(K).$$

Proof Let $\{\phi_n\}_{n=1}^\infty$ be a complete orthonormal basis of $\mathscr{H}_a(-)$ and $\{\psi_m\}_{m=1}^\infty$ that of $\mathscr{H}_a(+)$. We see that

$$0 \geq \mathrm{Tr}\, T\lceil_{\mathscr{H}_a(-)} = \sum_{n=1}^\infty (\phi_n, T\phi_n) \geq \sum_{n=1}^\infty (\phi_n, \tilde{T}\phi_n),$$

where $\tilde{T} = E^{\overline{h_{\rm eff}}}_{[0,|e_{\rm eff}|)} \otimes \left((\kappa^2 - \frac{\hat{\theta}_\kappa}{2})\hat{H}^\nu_{\rm rad}\lceil_{\mathscr{H}_a} - \hat{\theta}'_\kappa\right)$. This follows from Lemma 6.13: $T \geq \tilde{T}$. Set $\tilde{T}_- = \tilde{T}E^{\tilde{T}}_{(-\infty,0)}$. We have

$$0 \geq \mathrm{Tr}\ T\lceil_{\mathscr{H}_a(-)} \geq \sum_{n=1}^\infty (\phi_n, \tilde{T}_-\phi_n) \geq \sum_{n=1}^\infty (\phi_n, \tilde{T}_-\phi_n) + \sum_{m=1}^\infty (\psi_m, \tilde{T}_-\psi_m) = \mathrm{Tr}\tilde{T}_-.$$

Hence we obtain that

$$\left|\mathrm{Tr}\ T\lceil_{\mathscr{H}_a(-)}\right| \leq \left|\mathrm{Tr}\tilde{T}_-\right| = \mathrm{Tr}\, E^{\overline{h_{\rm eff}}}_{[0,|e_{\rm eff}|)} \times \left|\mathrm{Tr}\left((\kappa^2 - \frac{\hat{\theta}_\kappa}{2})\ \hat{H}^\nu_{\rm rad}\Big\lceil_{\mathscr{H}_a} - \hat{\theta}'_\kappa\right)_-\right|,$$

where $\mathrm{Tr}X_-$ denotes the sum of negative eigenvalues of X. Since $\mathrm{sp}(\hat{H}^\nu_{\rm rad}\Big\lceil_{\mathscr{H}_a}) = \mathrm{sp}_{disc}(\hat{H}^\nu_{\rm rad}\Big\lceil_{\mathscr{H}_a})$ and $\left|\mathrm{Tr}\,E^{\overline{h_{\rm eff}}}_{[0,|e_{\rm eff}|)}\right| < \infty$, it follows that $\left|\mathrm{Tr}\ T\lceil_{\mathscr{H}_a(-)}\right| < \infty$. Thus the lemma follows. ■

Lemma 6.15 *Suppose* $|\alpha| > \alpha_\varepsilon$ *and that* $\min\{|e_{\rm eff}|/3, 2\kappa^2\} > \theta_\kappa$. *Then for* ν *such that* $|e_{\rm eff}| > 3\theta_\kappa + \nu$, $\mathrm{sp}(H^\nu_\kappa) \cap [E(H^\nu_\kappa), E(H^\nu_\kappa) + \nu) \subset \mathrm{sp}_{\rm disc}(H^\nu_\kappa)$. *In particular* H^ν_κ *has the ground state.*

Proof Suppose $\min\{|e_{\rm eff}|/3, 2\kappa^2\} > \hat{\theta}_\kappa$. We see that $\mathrm{sp}(\hat{H}^\nu_\kappa) = \mathrm{sp}(K^\perp) \cup \mathrm{sp}(K)$, $\mathrm{sp}(K^\perp) \subset [E(K) + \nu, \infty)$ and $\mathrm{sp}(K) \cap [E(K), E(K) + \nu) \subset \mathrm{sp}_{\rm disc}(K)$. Notice that $E(K) = E(\hat{H}^\nu_\kappa)$. It follows that

$$\mathrm{sp}(\hat{H}^\nu_\kappa) \cap [E(\hat{H}^\nu_\kappa), E(\hat{H}^\nu_\kappa) + \nu) \subset \mathrm{sp}_{\rm disc}(\hat{H}^\nu_\kappa).$$

Note that $\lim_{L\to\infty}\lim_{a\to\infty} \hat{H}^\nu_\kappa = H^\nu_\kappa$ in the uniform resolvent sense. This completes the proof. ■

A normalized ground state of H^ν_κ is denoted by Ψ_ν.

Lemma 6.16 *Suppose* $|\alpha| > \alpha_\varepsilon$, *Assumption 6.10, and* $\min\{|e_{\rm eff}|/3, 2\kappa^2\} > \theta_\kappa$. *Then for* ν *such that* $|e_{\rm eff}| > 3\theta_\kappa + \nu$,

$$\|N^{1/2}\Psi_\nu\| \leq \frac{1}{\kappa}C\|\varrho/\omega^{5/2}\|(\max_\mu \|\nabla_\mu V\|_\infty)\|\Psi_\nu\| \tag{6.7}$$

with some constant C.

Proof We set $E = E(H^\nu_\kappa)$. In a similar way to the Pauli–Fierz Hamiltonian we can derive the pull-through formula:

$$\|N^{1/2}\Psi_\nu\|^2 = \frac{1}{2\kappa^2}\sum_{j=1,2}\int_{\mathbb{R}^3} \left\|(H^\nu_\kappa - E + \omega(k) + \nu)^{-1}T_\kappa^{-1}\nabla V \cdot \rho(k,j)T_\kappa\Psi_\nu\right\|^2 dk$$

We then have (6.7). ∎

Lemma 6.17 *Suppose $|\alpha| > \alpha_\varepsilon$ and Assumptions 6.10. Let P_Ω be the projection onto $\{\alpha\Omega \mid \alpha \in \mathbb{C}\}$ and $Q_\Omega = E^{\overline{h_{\rm eff}}}_{[\delta,\infty)} \otimes P_\Omega$ with some $\delta > \frac{3}{2}\theta_\kappa$. Suppose that $\min\{|e_{\rm eff}|/3, 2\kappa^2\} > \theta_\kappa$. Then for ν such that $|e_{\rm eff}| > 3\theta_\kappa + \nu$, it follows that*

$$\|Q_\Omega \Psi_\nu\| \le \sqrt{\frac{\theta_\kappa}{\delta - \frac{3}{2}\theta_\kappa}}\|\Psi_\nu\|.$$

Proof Since $(\Psi_\nu, Q_\Omega(H_\kappa^\nu - E(H_\kappa^\nu))\Psi_\nu) = 0$, we have

$$(\Psi_\nu, Q_\Omega(h_{\rm eff} - E(H_\kappa^\nu))\Psi_\nu) = -(\Psi_\nu, Q_\Omega \delta V_\kappa \Psi_\nu).$$

The left-hand side above is estimated as

$$(\Psi_\nu, Q_\Omega(h_{\rm eff} - E(H_\kappa^\nu))\Psi_\nu) \ge (e_{\rm eff} + \delta - E(H_\kappa^\nu))(\Psi_\nu, Q_\Omega\Psi_\nu).$$

Note that $e_{\rm eff} + \delta - E(H_\kappa^\nu) \ge \delta - \frac{3}{2}\theta_\kappa > 0$. We have

$$(\Psi_\nu, Q_\Omega(h_{\rm eff} - E(H_\kappa^\nu) + g)\Psi_\nu) \ge (\delta - \frac{3}{2}\theta_\kappa)\|Q_\Omega\Psi_\nu\|^2 > 0.$$

Moreover

$$|(\Psi_\nu, Q_\Omega\delta V_\kappa\Psi_\nu)| \le \theta_\kappa\left(\|H_{\rm rad}^{1/2}Q_\Omega\Psi_\nu\| + \|Q_\Omega\Psi_\nu\|\right)\|\Psi_\nu\| \le \theta_\kappa\|\Psi_\nu\|^2.$$

Hence we have $0 < (\delta - \frac{3}{2}\theta_\kappa)\|Q_\Omega\Psi_\nu\|^2 \le \theta_\kappa\|\Psi_\nu\|^2$ and the lemma follows. ∎

We normalize Ψ_ν, i.e., $\|\Psi_\nu\| = 1$. Take a subsequence ν' such that $\Psi_{\nu'}$ weakly converges to a vector $\Psi_{\rm g}$ as $\nu' \to \infty$.

Theorem 6.18 (Hiroshima and Spohn [38, Theorem 3.4]) *We suppose Assumption 6.10. Then for any $\varepsilon > 0$, there exists κ_ε such that for all $\kappa > \kappa_\varepsilon$, H_κ has the ground state for all α such that $|\alpha| > \alpha_\varepsilon$.*

Proof Let $E_\nu = E(H_\kappa^\nu)$ and $E = E(H_\kappa)$. We see that $\lim_{\nu\to 0} E_\nu = E$. By Proposition 3.34 it is sufficient to prove $\Psi_{\rm g} \ne 0$. Note that $F = E^{\overline{h_{\rm eff}}}_{[0,\delta)} \otimes P_\Omega$ is compact for sufficiently small $\delta > 0$. Note that $N + P_\Omega \ge \mathbb{1}$. Hence

$$\mathbb{1} \otimes N + \mathbb{1} \otimes P_\Omega = \mathbb{1} \otimes N + F + Q_\Omega \ge \mathbb{1},$$

and

$$F \ge \mathbb{1} - \mathbb{1} \otimes N - Q_\Omega.$$

Suppose that $\min\{|e_{\rm eff}|/3, 2\kappa^2\} > \theta_\kappa$ and $\delta > \frac{3}{2}\theta_\kappa$. Then for ν' such that $|e_{\rm eff}| > 3\theta_\kappa + \nu'$, we have

$$(\Psi_{\nu'}, F\Psi_{\nu'}) \geq 1 - \frac{1}{\kappa} C \|Q/\omega^{5/2}\| (\max_{\mu} \|\nabla_\mu V\|_\infty) - \frac{\theta_\kappa}{\delta - \frac{3}{2}\theta_\kappa}.$$

Note that $\sup_{\alpha\in\mathbb{R}} \|Q/\omega^{5/2}\| < \infty$ and $\lim_{\kappa\to\infty} \dfrac{\theta_\kappa}{\delta - \frac{3}{2}\theta_\kappa} = 0$ uniformly with respect to α. Hence for sufficiently large κ, $(\Psi_{\nu'}, F\Psi_{\nu'}) > \eta$ follows uniformly in ν' and α with some $\eta > 0$. Take $\nu' \to 0$ on both sides above. Since F is compact, we see that $F\Psi_{\nu'} \to F\Psi_g$ strongly and $(\Psi_g, F\Psi_g) > \eta$. In particular $\Psi_g \neq 0$, and Ψ_g is the ground state of H_κ. ∎

We can also show the existence of the ground state of the Pauli–Fierz Hamiltonian without a scaling parameter.

Corollary 6.19 *Let $\kappa = 1$, i.e., the Hamiltonian is not scaled. Suppose (1) and (2) of Assumption 6.10, and that $\lim_{\alpha\to\infty} \|Q/\omega^{n/2}\| = 0$ for $n = 3, 4, 5$. Then there exists $\alpha_* > \alpha_\varepsilon$ such that for all α with $|\alpha| > \alpha_*$, $H_{\rm PF}^{\rm dip}$ has the ground state.*

Proof We can see that $\theta_\kappa \to 0$ and $\|Q/\omega^{5/2}\| \to 0$ as $\alpha \to \infty$. Then for sufficiently large α with $\kappa = 1$, the massive ground state Ψ_ν exists. Furthermore we have

$$(\Psi_{\nu'}, F\Psi_{\nu'}) \geq 1 - C \|Q/\omega^{5/2}\| (\max_{\mu} \|\nabla_\mu V\|_\infty) - \frac{\theta_1}{\delta - \frac{3}{2}\theta_1},$$

where θ_1 is θ_κ with $\kappa = 1$. Since $\lim_{|\alpha|\to\infty} \|Q/\omega^{5/2}\| = 0$ and $\lim_{|\alpha|\to\infty} \frac{\theta_1}{\delta - \frac{3}{2}\theta_1} = 0$, we can conclude that $\Psi_g \neq 0$ for sufficiently large $|\alpha|$. Then the corollary follows. ∎

6.1.4 Transition From Unbinding to Binding

In the previous sections we showed the existence and absence of the ground state. Combining these results we can construct examples of the Pauli–Fierz Hamiltonian having transition from unbinding to binding according to the value of coupling constant α. See Fig. 6.1.

Lemma 6.20 *Suppose Assumption 6.4. Then H_κ has no ground state for all $\kappa > 0$ and all α such that $|\alpha| < \alpha_c$.*

Proof Define the unitary operator u_κ by $(u_\kappa f)(x) = k^{3/2} f(x/\kappa)$. We infer $V_\kappa = \kappa^{-2} u_\kappa V u_\kappa^{-1}$, $-\Delta = \kappa^{-2} u_\kappa (-\Delta) u_\kappa^{-1}$ and

$$\||V_\kappa|^{1/2}(-\Delta)^{-1}|V_\kappa|^{1/2}\| = \kappa^{-2}\|u_\kappa |V|^{1/2} u_\kappa^{-1} (-\Delta)^{-1} u_\kappa |V|^{1/2} u_\kappa^{-1}\| = \|K_0\|.$$

Fig. 6.1 Binding and unbinding

Hence the lemma follows from Theorem 6.9. ■

Theorem 6.21 (Hiroshima, Suzuki and Spohn [39, Theorem 4.2]) *Let arbitrary $\delta > 0$ be given. Suppose Assumptions 6.4 and 6.10. Then there exist an external potential $\tilde{V}$ and constants $\alpha_+ > \alpha_-$ such that*

(1) $0 < \alpha_+ - \alpha_- < \delta$,
(2) $H_{\rm PF}^{\rm dip}$ *has the ground state for* $|\alpha| > \alpha_+$ *but no ground state for* $|\alpha| < \alpha_-$.

Proof For $\delta > 0$ we take $\varepsilon > 0$ such that $\alpha_\varepsilon - \alpha_c < \delta$. Take a sufficiently large κ, and set $V_\kappa(x) = V(x/\kappa)/\kappa^2$. Define $H_{\rm PF}^{\rm dip}$ by the Pauli–Fierz Hamiltonian with potential V_κ. $H_{\rm PF}^{\rm dip}$ has the ground state for $|\alpha| > \alpha_\varepsilon$ by Theorem 6.18, and $H_{\rm PF}^{\rm dip}$ has no ground state for $|\alpha| < \alpha_c$ by Lemma 6.20. Set $\alpha_\varepsilon = \alpha_+$ and $\alpha_c = \alpha_-$. Then the proof is complete. ■

6.1.5 Enhanced Binding by Cutoff Functions

In this section we show the enhanced binding by cutoff functions, i.e., it can be shown that the Pauli–Fierz Hamiltonian with dipole approximation has the ground state if UV cutoff parameter Λ is sufficiently large.

We consider the cutoff function:

$$\hat{\varphi}(k) = \mathbb{1}_{[\lambda,\Lambda]}(k).$$

Thus $m_{\rm eff} = m + \frac{8}{3}\alpha^2\pi(\Lambda - \lambda)$. If Λ goes to infinity, then $m_{\rm eff}$ also goes to infinity. Intuitively we can expect for sufficiently large Λ, $H_{\rm PF}^{\rm dip}$ has the ground state. We have the corollary below.

Corollary 6.22 *Suppose Assumptions 6.4 and* $\Lambda < \frac{3}{8\pi}\alpha^{-2}(m_{\rm c} - m) + \lambda$. *Then* $H_{\rm PF}^{\rm dip}$ *has no ground state.*

Proof $\Lambda < \frac{3}{8\pi}\alpha^{-2}(m_{\rm c} - m) + \lambda$ implies that $m_{\rm eff} < m_{\rm c}$. Hence the corollary follows from Theorem 6.9. ■

We can also show the existence of the ground state for sufficiently large Λ.

Corollary 6.23 *Suppose (1) and (2) of Assumption 6.10. Then there exists Λ_* such that $H_{\rm PF}^{\rm dip}$ has the ground state for $\Lambda > \Lambda_*$.*

Proof We notice that

$$m_{\rm eff}(k) = m + \frac{8\pi\alpha^2}{3}\left\{(\Lambda-\lambda) - \frac{1}{2}\left(|k|\log\left|\frac{(|k|+\Lambda)(|k|-\lambda)}{(|k|+\lambda)(|k|-\Lambda)}\right| - i\pi\mathbb{1}_{[\lambda,\Lambda]}(|k|)\sqrt{|k|}\right)\right\}.$$

We have

$$\|Q/\omega^{n/2}\|^2 = \int_{\lambda \le |k| \le \Lambda} \frac{1}{|m_{\mathrm{eff}}(k)|^2 \omega(k)^n} dk$$

and

$$\mathbb{1}_{[\lambda,\Lambda]}(k) \frac{1}{|m_{\mathrm{eff}}(k)|^2 \omega(k)^n} \le \left(\frac{3}{4\pi^2}\right)^2 \frac{1}{\omega(k)^{n+1}} \mathbb{1}_{[\lambda,\Lambda]}(k).$$

Since the right-hand side above is integrable for $n = 3, 4$ and 5. The the Lebesgue dominated convergence theorem yields that $\lim_{\Lambda\to\infty} \|Q/\omega^{n/2}\| = 0$. Hence in a similar way to Corollary 6.19 we can prove the corollary. ∎

6.2 Enhanced Binding for the Nelson Model

6.2.1 *The N-Body Nelson Hamiltonian*

In this section we consider the enhanced binding for the N-body Nelson Hamiltonian. The N-body Nelson Hamiltonian describes a linear interaction between N-non-relativistic spinless nucleons and a scalar meson field.

Definition 6.24 (*N-body Nelson Hamiltonian*) The N-body Nelson Hamiltonian is defined by

$$H_{\mathrm{N}} = \left(-\sum_{j=1}^N \frac{1}{2m_j}\Delta_j + V\right) \otimes \mathbb{1} + \mathbb{1} \otimes H_{\mathrm{f}} + \sum_{j=1}^N g_j \phi_j \tag{6.8}$$

acting on $L^2(\mathbb{R}^{dN}) \otimes \mathscr{F}$, where $g_j \in \mathbb{R}$, $j = 1, \dots, N$, are coupling constants and

$$\phi_j = \int_{\mathbb{R}^{dN}}^{\oplus} \phi(x_j) dx.$$

We consider external potentials of the form

$$V(x_1, \dots x_N) = \sum_{j=1}^N V_j(x_j).$$

Since there is no interaction between different particles, the jth particle is governed only by potential V_j. In this case, if V_j's are negative and sufficiently shallow, external potential $\sum_{j=1}^N V_j$ can not trap these particles. However if these particles attractively interact with each other by an effective potential provided by the scalar quantum field, particles close up with each other and seem to behave like as a single particle but with heavy mass $\sum_{j=1}^N m_j$. Hence H_N has the ground state. See Fig. 1.6.

We introduce assumptions:

Assumption 6.25 (Cutoff functions) For all $j = 1, ..., N$, $\varphi_j \in \mathscr{S}'(\mathbb{R}^d)$ satisfies that

(1) $\hat{\varphi}_j \in L^1_{\mathrm{loc}}(\mathbb{R}^d)$,
(2) $\hat{\varphi}_j(-k) = \overline{\hat{\varphi}_j(k)}$ and $\hat{\varphi}_j/\sqrt{\omega}, \hat{\varphi}_j/\omega \in L^2(\mathbb{R}^d)$.
(3) $\hat{\varphi}_j \in C_0^2(\mathbb{R}^d)$.
(4) For all $p \in [1, 2)$ and $\hat{\varphi}_j \in W^{1,p}(\mathbb{R}^d)$.

This assumption is assumed in this section unless otherwise stated. Suppose that $\int_{\mathbb{R}^d} \hat{\varphi}_j^2(k)/\omega(k)^3 dk < \infty$, $j = 1, ..., N$, and define the Gross transformation e^T on $\mathscr{H}$ by

$$T = \exp\left(-i\sum_{j=1}^N g_j \pi_j\right),$$

where $\pi_j = \displaystyle\int_{\mathbb{R}^{dN}}^{\oplus} \pi_j(x_j) dx$ with

$$\pi_j(x) = \frac{i}{\sqrt{2}} \int \left(a^\dagger(k)e^{-ikx}\frac{\hat{\varphi}_j(k)}{\omega(k)^{3/2}} - a(k)e^{ikx}\frac{\hat{\varphi}_j(-k)}{\omega(k)^{3/2}}\right) dk.$$

We can show that T maps $D(H_\mathrm{N})$ onto itself and

$$T^{-1} H_\mathrm{N} T = H_{\mathrm{eff}} + H_\mathrm{f} + H_\mathrm{I}. \tag{6.9}$$

Here H_{eff} is an effective Hamiltonian given by

$$H_{\mathrm{eff}} = \sum_{j=1}^N \left(-\frac{1}{2m_j}\Delta_j + V_j\right) + V_{\mathrm{eff}},$$

with an effective potential

$$V_{\mathrm{eff}}(x) = -\frac{1}{4}\sum_{i \neq j}^N g_i g_j \int_{\mathbb{R}^d} \frac{\hat{\varphi}_i(-k)\hat{\varphi}_j(k)}{\omega(k)^2} e^{-ik(x_i - x_j)} dk.$$

The remainder term in (6.9) is given by

$$H_\mathrm{I} = \sum_{j=1}^N \left\{\frac{g_j}{2m_j}(p_j A_j + A_j p_j) + \frac{g_j^2}{2m_j} A_j^2 - \frac{g_j^2}{2}\|\hat{\varphi}_j/\sqrt{\omega}\|^2\right\},$$

where $p_j = -i\nabla_j$ and $A_j = \int_{\mathbb{R}^{dN}}^{\oplus} A_j(x_j)dx$ with

$$A_j(x) = \frac{1}{\sqrt{2}} \int k \left(a^\dagger(k)e^{-ikx}\frac{\hat{\varphi}_j(-k)}{\omega(k)} + a(k)e^{ikx}\frac{\hat{\varphi}_j(k)}{\omega(k)} \right) dk.$$

An assumption is introduced:

Assumption 6.26 (1) There exists $g_c > 0$ such that $\inf \mathrm{sp}(H_{\mathrm{eff}}) \in \mathrm{sp}_{\mathrm{disc}}(H_{\mathrm{eff}})$ for g_j with $|g_j| > g_c$, $j = 1, ..., N$.
(2) $V_j(-\Delta_j + \mathbb{1})^{-1}$, $j = 1, \ldots, N$, are compact.

Let us set $[N] = \{1, ..., N\}$. For $\beta \subset [N]$ we set $|\beta| = \#\beta$. For $\beta \subset [N]$, we define

$$H_0(\beta) = \sum_{j\in\beta} \frac{1}{2m_j} \left(-i\nabla_j - g_j A_j\right)^2 + H_f + V_{\mathrm{eff}}(\beta),$$

$$V_{\mathrm{eff}}(\beta) = \begin{cases} -\frac{1}{4}\sum_{i,j\in\beta, i\neq j} g_i g_j \int_{\mathbb{R}^d} \frac{\hat{\varphi}_i(-k)\hat{\varphi}_j(k)}{\omega(k)^2} e^{-ik(x_i - x_j)}dk, & |\beta| \geq 2, \\ 0, & |\beta| = 0, 1. \end{cases}$$

We set

$$H_V(\beta) = H_0(\beta) + \sum_{j\in\beta} V_j.$$

Simply we set $H_V = H_V([N])$. The operators $H_0(\beta)$ and $H_V(\beta)$ are selfadjoint operators acting on $L^2(\mathbb{R}^{d|\beta|}) \otimes \mathscr{F}$. We set

$$E_V = \inf \mathrm{sp}(H_V), \quad E_V(\beta) = \inf \mathrm{sp}(H_V(\beta)),$$
$$E_0(\beta) = \inf \mathrm{sp}(H_0(\beta)), \quad E_V(\emptyset) = 0.$$

The lowest two cluster threshold Σ_V is defined by

$$\Sigma_V = \min\{E_V(\beta) + E_0(\beta^c) | \beta \subsetneq [N]\}.$$

6.2.2 *Existence of the Ground State*

In Chap. 3 we show the existence of the ground state of the Pauli–Fierz Hamiltonian. Binding condition, Assumption 3.10, is a sufficient condition for the existence of the ground state in the massive case. By a limiting argument we can also see the existence of the ground state in the massless case. Similarly to establish the existence of the ground state of H_{N}, we use the next proposition:

Proposition 6.27 (Binding condition) *Suppose Assumption 6.26 and $\Sigma_V - E_V > 0$. Then H_{N} has the ground state.*

By $\Sigma_V - E_V > 0$ we can show the existence of the ground state in the massive case. By the same limiting argument as that of the Pauli–Fierz Hamiltonian we can also see the existence of the ground state in the massless case. For $\beta \subset [N]$, we set the Schrödinger operators in $L^2(\mathbb{R}^{d|\beta|})$ by

$$h_0(\beta) = -\sum_{j\in\beta} \frac{1}{2m_j}\Delta_j + V_{\text{eff}}(\beta), \quad h_V(\beta) = h_0(\beta) + \sum_{j\in\beta} V_j,$$
$$\mathscr{E}_0(\beta) = \inf \operatorname{sp}(h_0(\beta)), \quad \mathscr{E}_V(\beta) = \inf \operatorname{sp}(h_V(\beta)),$$

where $h_0(\emptyset) = 0$ and $h_V(\emptyset) = 0$. Furthermore we simply put $h_V = h_V([N]) = H_{\text{eff}}$ and $\mathscr{E}_V = \inf \operatorname{sp}(h_V)$. We define the lowest two cluster threshold for h_V by

$$\Xi_V = \min\{\mathscr{E}_V(\beta) + \mathscr{E}_0(\beta^c) | \beta \subsetneqq [N]\}$$

and we set

$$V_{\text{eff}}^{ij}(x) = -\frac{1}{4} g_i g_j \int_{\mathbb{R}^d} \frac{\hat{\varphi}_i(-k)\hat{\varphi}_j(k)}{\omega(k)^2} e^{-ikx} dk, \quad i \neq j.$$

We introduce a scaling for H_{N}. We define H_κ by

$$H_\kappa = \left(-\sum_{j=1}^N \frac{1}{2m_j}\Delta_j + V\right) + \kappa^2 H_{\text{f}} + \kappa \sum_{j=1}^N g_j \phi_j.$$

This scaling can be introduced by replacing $a^\sharp$ with $\kappa a^\sharp$ or $\hat{\varphi}_j$ and ω with $\kappa\hat{\varphi}_j$ and $\kappa^2\omega$, respectively. Hence Σ_V and E_V depend on κ, and we rewrite them as $\Sigma_V(\kappa)$ and $E_V(\kappa)$, respectively. The scaling parameter κ can be regarded as a dummy and absorbed into m_j, V_j and $\hat{\varphi}_j$, $j = 1, \ldots, N$. Define

$$\hat{H} = \sum_{j=1}^N \left(-\frac{1}{2\hat{m}_j}\Delta_j + \hat{V}_j\right) + \sum_{j=1}^N g_j \hat{\phi}_j + H_{\text{f}},$$

where $\hat{m}_j = m_j\kappa^2$, $\hat{V}_j = V_j/\kappa^2$ and $\hat{\phi}_j$ is defined by ϕ_j with $\hat{\varphi}_j$ replaced by $\hat{\varphi}_j/\kappa$. We have $\kappa^{-2}H_\kappa = \hat{H}$. If H_κ has a ground, then $\hat{H}$ also has a ground state. To show the existence of the ground state of $\hat{H}$ it is sufficient to show $\Sigma_V(\kappa) - E_V(\kappa) > 0$ for some κ by Proposition 6.27.

Lemma 6.28 *For any* $\beta \subset [N]$, *it follows that* $\inf \operatorname{sp}(h_V(\beta)) \leq \inf \operatorname{sp}(H_V(\beta))$.

Proof This is a consequence of the diamagnetic inequality. See [30]. ■

Lemma 6.29 *(1) For an arbitrary* $\kappa > 0$, *it follows that* $\Sigma_V(\kappa) \geq \Xi_V$.
(2) Assume (1) of Assumption 6.26. Then

$$E(\kappa) \leq \mathscr{E}_V + \kappa^{-2} \frac{1}{4m_j} \sum_{j=1}^{N} g_j^2 \|\hat{\varphi}_j\|^2$$

for g_j *with* $|g_j| > g_c$, $j = 1, ..., N$.

Proof (1) follows from Lemma 6.28, the definition of the lowest two cluster thresholds and the fact $\Xi_V = \inf \mathrm{sp}_{\mathrm{ess}}(H_{\mathrm{eff}})$. By (1) of Assumption 6.26, H_{eff} has the normalized ground state u for g_j with $|g_j| > g_c$, $j = 1, ..., N$. Set $\Psi = u \otimes \Omega$. We have

$$E(\kappa) \leq (\Psi, H_\kappa \Psi) \leq \mathscr{E}_V + \sum_{j=1}^{N} \frac{g_j^2}{4m_j \kappa^2} \|\hat{\varphi}_j\|^2.$$

Here we used that $(\nabla_j \Psi, A_j \Psi) = 0$. Then (2) follows. ■

Theorem 6.30 (Hiroshima and Sasaki [37, Theorem 2.3]) *We suppose Assumption 6.26 and* $\int_{\mathbb{R}^d} \hat{\varphi}_j^2(k)/\omega(k)^3 dk < \infty$, $j = 1, ..., N$. *Fix a sufficiently large* $\kappa > 0$. *Then there exists* $g_c(\kappa)$ *such that for* g_j *with* $g_c < |g_j| < g_c(\kappa)$, $j = 1, ..., N$, H_κ *has the ground state, where* $g_c(\kappa)$ *is possibly infinity.*

Proof By Lemmas 6.29, we have

$$\Sigma_V(\kappa) - E(\kappa) \geq \Xi_V - \mathscr{E}_V - \sum_{j=1}^{N} \frac{g_j^2}{4m_j \kappa^2} \|\hat{\varphi}_j\|^2$$

Note that $\Xi_V - \mathscr{E}_V > 0$ and $\Xi_V - \mathscr{E}_V$ is continuous in $g_1, ..., g_N$. For a sufficiently large κ, there exists $g_c(\kappa) > g_c$ such that $\Sigma_V(\kappa) - E(\kappa) > 0$ for $g_c < |g_j| < g_c(\kappa)$, $j = 1, ..., N$. Thus H_κ has the ground state for such g_j's by Proposition 6.27. ■

Corollary 6.31 *We suppose Assumption 6.26 and* $\int_{\mathbb{R}^d} \hat{\varphi}_j^2(k)/\omega(k)^3 dk < \infty$, $j = 1, ..., N$. *Then* $\hat{H}$ *has the ground state for* $g_c < |g_j| < g_c(\kappa)$, $j = 1, ..., N$, *where* $g_c(\kappa)$ *is introduced in Theorem 6.30.*

Proof We have $\kappa^{-2} H_\kappa = \hat{H}$. By Theorem 6.30, $\hat{H}$ has the ground state. ■

We show a typical example of cutoff functions and effective potentials. Let $\hat{\varphi}_j = \rho_j/\sqrt{\omega}$, $j = 1, ..., N$, with rotation invariant nonnegative functions ρ_j. In this case, effective potential V_{eff} is explicitly computed as

$$V_{\mathrm{eff}}(x_1, \ldots, x_N) = -\frac{1}{4} \sum_{i \neq j}^{N} g_i g_j \frac{\sqrt{(2\pi)^d}}{|x_i - x_j|^{(d-1)/2}} \int_0^\infty \frac{r^{(d-1)/2}}{r^2} \rho_i(r) \rho_j(r) \sqrt{r|x_i - x_j|} J_{\frac{d-2}{2}}(r|x|) dr.$$

Here J_ν is the Bessel function: $J_\nu(x) = (\frac{x}{2})^\nu \sum_{n=0}^{\infty} \frac{(-1)^n}{n!\Gamma(n+\nu+1)} (\frac{x}{2})^{2n}$. We can see that V_{eff} satisfies that

(1) $V_{\rm eff}^{ij}$ is continuous,
(2) $\lim_{|x|\to\infty} V_{ij}(x) = 0$,
(3) $V_{\rm eff}^{ij}(0) < V_{\rm eff}^{ij}(x)$ for all $x \in \mathbb{R}^d$ but $x \neq 0$.

We give an example of $V_1, \ldots, V_N$ satisfying (1) of Assumption 6.26. Assume simply that $V_1 = \ldots = V_N = V$, $g_1 = \ldots = g_N = g$, $\hat{\varphi}_1 = \ldots = \hat{\varphi}_N = \lambda$ and $m_1 = \ldots = m_N = m$. Then

$$V_{\rm eff}^{ij}(x) = W(x) = -\frac{g^2}{4}\int_{\mathbb{R}^d} \frac{|\lambda(k)|^2}{\omega(k)^2} e^{-ikx} dk \tag{6.10}$$

for all $i \neq j$. Let

$$h_V(g) = \sum_{j=1}^N \left(-\frac{1}{2m}\Delta_j + V(x_j)\right) + g^2 \sum_{j\neq l}^N W(x_j - x_l),$$

which acts on $L^2(\mathbb{R}^{dN})$.

Assumption 6.32 We suppose (1), (2) and (3):

(1) V is relatively compact with respect to Δ, and $\mathrm{sp}(-(\Delta/2m) + V) = [0, \infty)$.
(2) W satisfies that $-\infty < W(0) = \mathrm{essinf}_{|x|<\epsilon} W(x) < \mathrm{essinf}_{|x|>\epsilon} W(x)$ for all $\epsilon > 0$.
(3) $\inf \mathrm{sp}(-(\Delta/(2Nm) + NV) \in \mathrm{sp}_{\rm disc}(-(\Delta/(2Nm) + NV)$.

W given by (6.10) satisfies (2), and that $\lim_{|x|\to\infty} W(x) = 0$ and $W(x)$ is relatively compact with respect to $-\Delta$. The condition (1) means that the external potential V is shallow and the non-interacting Hamiltonian $h_V(0)$ has no negative energy bound state. We can prove the theorem below.

Theorem 6.33 (Hiroshima and Sasaki [37, Theorem 3.5]) *We suppose Assumption 6.32. Then there exists $g_c > 0$ such that for all g with $|g| > g_c$,*

$$\inf \mathrm{sp}(h_V(g)) \in \mathrm{sp}_{\rm disc}(h_V(g)).$$

Namely $h_V(g)$ for $|g| > g_c$ has the ground state.

References

1. W.O. Amrein, *Hilbert Space Methods in Quantum Mechanics* (EPFL Press, Lausanne, 2009)
2. A. Arai, Rigorous theory of spectra and radiation for a model in quantum electrodynamics. J. Math. Phys. **24**, 1896–1910 (1983)
3. A. Arai, An asymptotic analysis and its applications to the nonrelativistic limit of the Pauli-Fierz and a spin-boson model. J. Math. Phys. **31**, 2653–2663 (1990)
4. A. Arai, *Analysis of Fock Spaces and Mathematical Theory of Quantum Fields* (World Scientific, Singapore, 2018)
5. A. Arai, M. Hirokawa, On the existence and uniqueness of ground states of a generalized spin-boson model. J. Funct. Anal. **151**, 455–503 (1997)
6. V. Bach, J. Fröhlich, I.M. Sigal, Mathematical theory of nonrelativistic matter and radiation. Lett. Math. Phys. **34**, 183–201 (1995)
7. V. Bach, J. Fröhlich, I.M. Sigal, Quantum electrodynamics of confined nonrelativistic particles. Adv. Math. **137**, 299–395 (1998)
8. V. Bach, J. Fröhlich, I.M. Sigal, Renormalization group analysis of spectral problems in quantum field theory. Adv. Math. **137**, 205–298 (1998)
9. V. Bach, J. Fröhlich, I.M. Sigal, Spectral analysis for systems of atoms and molecules coupled to the quantized radiation field. Commun. Math. Phys. **207**, 249–290 (1999)
10. M. Born, W. Heisenberg, P. Jordan, Zur Quantenmechanik. II. Zeitschrift für Physik **35**, 557–615 (1926)
11. H. Brezis, *Functional Analysis, Sobolev Spaces and Partial Differential Equations* (Springer, Berlin, 2011)
12. R. Carmona, Pointwise bounds for Schrödinger eigenstates. Commun. Math. Phys. **62**, 97–106 (1978)
13. J. Dereziński, C. Gérard, Scattering theory of infrared divergent Pauli-Fierz Hamiltonians. Ann. Henri Poincaré **5**, 523–578 (2004)
14. L.C. Evans, *Partial Differential Equations* (American Mathematical Society, Providence, 2002)
15. M. Falconi, Self-adjointness criterion for operators in Fock spaces. Math. Phys. Anal. Geom. **18**(18) (2015)
16. E.I. Fredholm, Sur une classe déquations fonctionnelles. Acta Math. **27**, 365–390 (1903)
17. C. Gérard, On the existence of ground states for massless Pauli-Fierz Hamiltonians. Ann. H. Poincaré **1**, 443–459 (2000)
18. C. Gérard, A remark on the paper: "On the existence of ground states for Hamiltonians". mp-arc 06-146, preprint (2006)

F. Hiroshima, *Ground States of Quantum Field Models*,
SpringerBriefs in Mathematical Physics,
https://doi.org/10.1007/978-981-32-9305-2

19. J. Glimm, A. Jaffe, A $\lambda\phi^4$ quantum field theory without cutoffs. I. Phys. Rev. **176**, 1945–1951 (1968)
20. M. Griesemer, Exponential decay and ionization thresholds in non-relativistic quantum electrodynamics. J. Funct. Anal. **210**, 321–340 (2004)
21. M. Griesemer, E. Lieb, M. Loss, Ground states in non-relativistic quantum electrodynamics. Invent. Math. **145**, 557–595 (2001)
22. S.J. Gustafson, I.M. Sigal, *Mathematical Concepts of Quantum Mechanics*, 2nd edn. (Springer, Berlin, 2011)
23. D. Hasler, I. Herbst, On the self-adjointness and domain of Pauli-Fierz type Hamiltonians. Rev. Math. Phys. **20**, 787–800 (2008)
24. D. Hasler, I. Herbst, Ground states in the spin boson model. Ann. Henri Poincaré **12** (2011)
25. W. Heisenberg, Über quantentheoretische Umdeutung kinematischer und mechanischer Beziehungen. Zeitschrift für Physik **33**, 879–893 (1925)
26. B. Helffer, J. Sjöstrand, Equation de Schrödinger avec champ magnétique et équation de Harper, in *Lecture Notes in Physics*, vol. 345 (Springer, Berlin, 1989), pp. 118–197
27. M. Hirokawa, Infrared catastrophe for Nelson's model-non-existence of ground state and soft-boson divergence. Publ. Res. Inst. Math. Sci. **42**, 897–922 (2006)
28. M. Hirokawa, F. Hiroshima, J. Lőrinczi, Spin-boson model through a Poisson driven stochastic process. Math. Zeitschrift **277**, 1165–1198 (2014)
29. M. Hirokawa, F. Hiroshima, H. Spohn, Ground state for point particles interacting through a massless scalar bose field. Adv. Math. **191**, 339–392 (2005)
30. F. Hiroshima, Diamagnetic inequalities for systems of nonrelativistic particles with a quantized field. Rev. Math. Phys. **8**, 185–203 (1996)
31. F. Hiroshima, Functional integral representation of a model in quantum electrodynamics. Rev. Math. Phys. **9**, 489–530 (1997)
32. F. Hiroshima, Essential self-adjointness of translation-invariant quantum field models for arbitrary coupling constants. Commun. Math. Phys. **211**, 585–613 (2000)
33. F. Hiroshima, Observable effects and parametrized scaling limits of a model in nonrelativistic quantum electrodynamics. J. Math. Phys. **43**, 1755–1795 (2002)
34. F. Hiroshima, Self-adjointness of the Pauli-Fierz Hamiltonian for arbitrary values of coupling constants. Ann. Henri Poincaré **3**, 171–201 (2002)
35. F. Hiroshima, Multiplicity of ground states in quantum field models: applications of asymptotic fields. J. Funct. Anal. **224**, 431–470 (2005)
36. F. Hiroshima, O. Matte, Ground states and their associated Gibbs measures in the renormalized Nelson model (2019), arXiv:1903.12024
37. F. Hiroshima, I. Sasaki, Enhanced binding for N particle system interacting with a scalar field I. Math. Z. **259**, 657–680 (2007)
38. F. Hiroshima, H. Spohn, Enhanced binding through coupling to a quantum field. Ann. Henri Poincaré **2**, 1159–1187 (2001)
39. F. Hiroshima, A. Suzuki, H. Spohn, The non-binding regime of the Pauli-Fierz model. J. Math. Phys. **52**(062104), 12 (2011)
40. T. Kato, *Perturbation Theory for Linear Operators* (Springer, Berlin, 1966)
41. A.J. Leggett, S. Chakravarty, A.T. Dorsey, M.P.A. Fisher, A. Garg, W. Zwerger, Dynamics of the dissipative two-state system. Rev. Mod. Phys. **59**, 1–85 (1987)
42. E. Lieb, Bounds on the eigenvalues of the Laplacian and Schrödinger operator. Bull. Am. Math. Soc. **82**, 751–753 (1976)
43. E. Lieb, M. Loss, Existence of atoms and molecules in non-relativistic quantum electrodynamics. Adv. Theor. Math. Phys. **7**, 667–710 (2003)
44. E. Lieb, W.E. Thirring, Bound for the kinetic energy of fermions which prove the stability of matters. Phys. Rev. Lett. **35**, 687–689 (1975)
45. O. Matte, Pauli-Fierz type operators with singular electromagnetic potentials on general domains. Math. Phys. Anal. Geom. **20**, 41 (2017)
46. O. Matte, J. Møller, Feynman-Kac formulas for the ultra-violet renormalized Nelson model (2017), arXiv:1701.02600

47. S. Molchanov, B. Vainberg, On general Cwikel-Lieb-Rozenblum and Lieb-Thirring inequalities (2012), arXiv:0812.2968v4
48. E. Nelson, Analytic vectors. Ann. Math. **70**, 572–615 (1959)
49. E. Nelson, Interaction of nonrelativistic particles with a quantized scalar field. J. Math. Phys. **5**, 1990–1997 (1964)
50. W. Pauli, M. Fierz, Zur Theorie der Emission langwelliger Lichtquanten. Nuovo Cimento **15**, 167–188 (1938)
51. M. Reed, B. Simon, *Methods of Modern Mathematical Physics II* (Academic, New York, 1975)
52. M. Reed, B. Simon, *Methods of Modern Mathematical Physics I* (Academic Press, New York, 1980)
53. B. Simon, *Trace Ideals and Their Applications*, 2nd edn. (American Mathematical Society, Providence, 2005)
54. H. Spohn, Ground state of quantum particle coupled to a scalar boson field. Lett. Math. Phys. **44**, 9–16 (1998)
55. H. Spohn, *Dynamics of Charged Particles and Their Radiation Field* (Cambridge University Press, Cambridge, 2004)
56. J. Weidmann, *Linear Operators in Hilbert Spaces* (Springer, Berlin, 1980)

Index

F. Hiroshima, *Ground States of Quantum Field Models*,
SpringerBriefs in Mathematical Physics,
https://doi.org/10.1007/978-981-32-9305-2

Printed in the United States
By Bookmasters